Paradise Rescu

It's Not About the Dirt

David Stannard

Dedication

You have provided my Mission,
painted my Vision and
empowered my Passion.
Thank you Cardan, Bordeaux
and everyone who has connected
with project Paradise Rescued
across the world.

Failure is not an option.

ISBN: 978-0-6485962-4-0

Published by Maria Carlton Pty Ltd, QLD, Australia
www.mariacarlton.com

Design and Photography by Sweet Graphic Design, QLD Australia
www.sweetgraphicdesign.com.au

Printed by Create Space, An Amazon.com company

Contents

SAS PARADISE RESCUED - VISION 2020

Mission Vision Passion

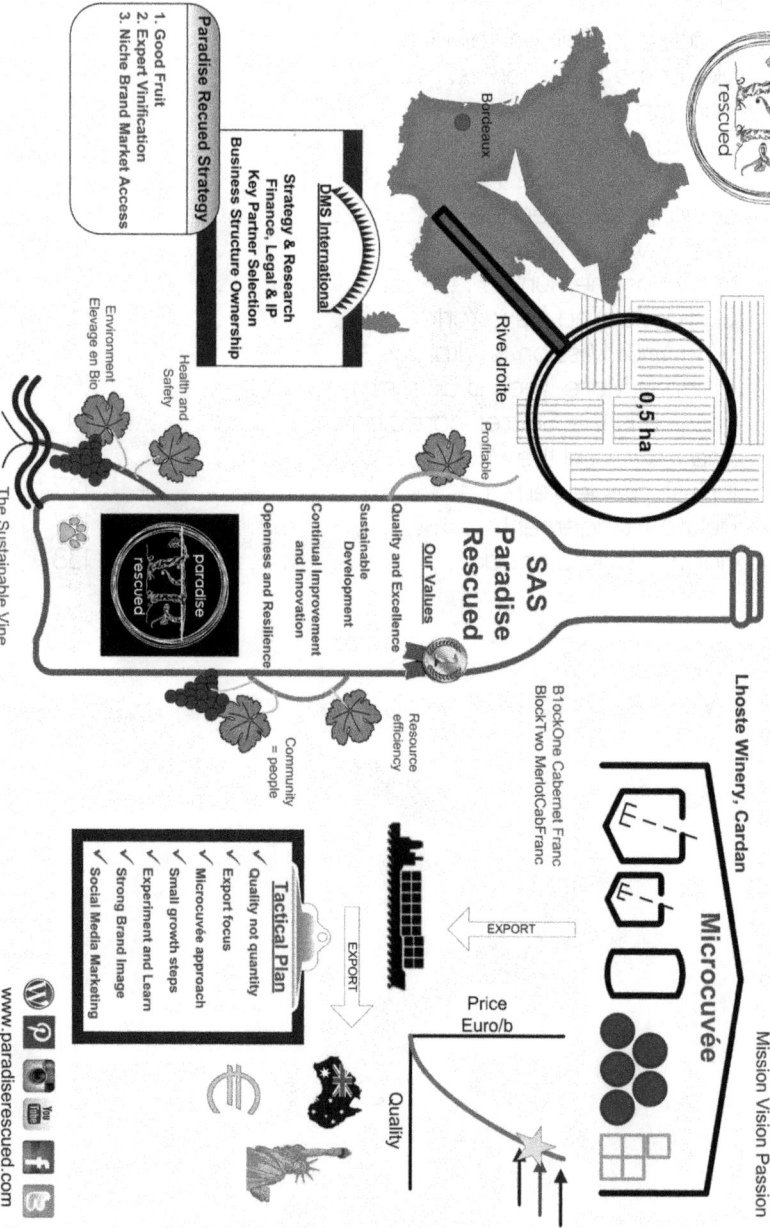

paradise rescued

Bordeaux

Rive droite

0.5 ha

Lhoste Winery, Cardan

Paradise Recued Strategy

1. Good Fruit
2. Expert Vinification
3. Niche Brand Market Access

DMS International

Strategy & Research
Finance, Legal & IP
Key Partner Selection
Business Structure Ownership

Health and Safety

Environment
Elevage en Bio

The Sustainable Vine

Profitable

SAS Paradise Rescued

Our Values

Quality and Excellence

Sustainable Development

Continual Improvement and Innovation

Openness and Resilience

paradise rescued

Resource efficiency

Community = people

BlockOne Cabernet Franc
BlockTwo MerlotCabFranc

Microcuvée

EXPORT

EXPORT

Tactical Plan

✓ Quality not quantity
✓ Export focus
✓ Microcuvée approach
✓ Small growth steps
✓ Experiment and Learn
✓ Strong Brand Image
✓ Social Media Marketing

Price Euro/b

Quality

www.paradiserescued.com

Introduction

I have come to believe that if we, as human beings, are not growing then we are probably starting to rot. Human personal development is critical to our existence and when it stops, we go backwards. Pretty much the same can be said about a business too.

During the time it has taken me to write this second book, a significant number of businesses will have failed. The reasons are many and varied and our experience will not be the same journey as yours. But we don't want you to 'trial' the experience to prove us right!

Thankfully Paradise Rescued has moved on – we have passed our sixth birthday. And our mission continues strongly, if not more resolutely than ever. I write this not as a statement of arrogance, but self-recognition that a small team of talented passionate dedicated people are continuing to reach new heights in excellence and personal achievement in pursuit of a unique cause and objective – our mission and vision.

There have been many highs and there have also been many lows. It is easy to look at our social media and assume that "Cabernet Franc" is growing everywhere and the challenge of sustaining a rural heritage, community vineyard and way of life has been trouble free. We clearly understand that 'bad news sells' but that's not our style. We recognise that many lessons have had to be learnt and no matter how much wisdom one can seek and accumulate, the path to personal and business growth has to be experienced.

In "From Cabbage Patch to Cabernet Franc" I shared my story covering the events that led to Paradise Rescued coming into existence and its action packed and challenging first year.

The purpose of this book is to not only share some more of our journey but also to focus on helping readers to benefit from our mistakes. We are open in our admission of what happened, where it wasn't successful and what we then learned from it. If we can at least help a few other people or businesses avoid the errors we made, then this will be a positive outcome.

There is also a human theme. Businesses are run and operated by people for people in some way or other. Many of the lessons here are personal as well as commercial. There was (and still is) a process of my self-learning, personal development and leadership growth that I needed to undertake with Paradise Rescued in order that our success could be achieved. Although born into an entrepreneurial family, I have spent much of my life as a management engineer in a completely different commercial environment, in terms of both the nature of the business and its' scale. Pretty much all of the challenges were of my making or lack of experience. Much of the learning is about my personal journey as a leader and what I had to learn to make Paradise Rescued succeed. I first had to grow in order that my business could then grow.

This book is still a storybook, but with a difference. Sharing our journey and experience gives it life, context and reality. The learnings contained within are not theoretical. They were all learnt during the journey either by good planning, better listening or hard real experience. I hope you can relate, but do not repeat.

I will take a detailed look at some of the key things that my team and I found to be critical in steadily building a sustainable, small winery business and brand that I believe will be useful to your business too. At the end of each short chapter, I share what I believe were my key learnings. I challenge every reader to make your own notes and set a goal for short-term action, change and improvement.

Not everything will apply to other small medium enterprises or even big corporations – a wine business is after all just that – a wine business. However, after six years covering many failures and also many better moments, we think we have some really good lessons to share.

I like the idea of the story continuing! Why? Quite simply because if it wasn't continuing, changing and adding new dimensions, then neither I personally, Team Paradise Rescued, nor our business would be changing either. Even a model of sustainability needs to change, adapt and grow. Just like a human being? We must never forget that leadership is centered around people, not things.

I want to see and hear your brand stories too. I want to sit and listen to your inspiration, journey and leadership. But, please remember, the failures are easily forgotten. In order to tell your story, you first have to walk the walk to success.

I hope that in another few years from now we will be sharing even more of our story - From Cabbage Patch to World Wide Brand! And of course our Mission to sustain the ongoing heritage and culture of our community will be even stronger. But I also know that by then I will have learned further lessons to share.

Grow yourself. Grow your business.

David Stannard

Chapter 1

Success is Built on Failures

Despite our first vintage in 2010 being an outstanding early success, 2011 was an unmitigated disaster. It failed, we failed – big time! We got it wrong. There are no Paradise Rescued labels with 2011 written on them as our vintage.

This is no emotional sympathy seeking first chapter trying to buy sales. But I hope it is attention grabbing? Why? Because I see so many organisations and people around the world just covering up their mistakes to protect corporate and personal egos rather than learning from the opportunities offered from their mistakes.

My team has always begged me not to write about nor talk about the Paradise Rescued vintage 2011. "We just can't do that sort of thing in Bordeaux". "Our reputation will be destroyed". "Best keep quiet and it will go away!" And so the crazy lines go on. But, I disagree.

Certainly, it may not be a good idea to hold a press conference to announce your failure, but it is a much worse idea to hope that the world doesn't find out about it and then suffer the significantly worse consequences when it finally does and the story is all over the media before you have decided to 'come clean' on the issue!

I understand that our style of thinking, operation and communication are possibly a touch different. Many would say weird and unrealistically open. However, for us, a big part of being sustainable is telling our story as it is or was. If no one is going to be specifically hurt in the process, then why not?

It's natural to think that it's easy to "fake it 'till you make it" and in so doing, to cover up our mistakes and pretend they never happened. And then we all "look good" when the internal and external world does its analyses?

However the issues always show up somewhere, sometime later so it is far better to just be open and honest up front. Hiding reality is a false ego saving game that serves no real purpose and effectively prevents an organisation from taking the necessary steps to growing, improving and moving forward. In order to focus on excellence, the dialogue must be open and strong.

When it goes wrong, it is actually better to admit it up front, change and move forward in order to stop that mistake happening again. Denial of an issue makes one very vulnerable to it happening again! It won't go away unless you deal with it.

That Heart Stopping Call

Vintage 2011 was a pivotal moment in our journey that has contributed enormously to the quality of our wine and winery operations since. We are NOT proud of the event but happy that we had it at that moment back then and have learnt significantly from it. We are now using it to our advantage!

I clearly remember the telephone call. It was mid October 2011. I am very proud of our French Vineyard Operations team for how they managed every aspect of the matter. It was late Tuesday evening in Melbourne as I took the call from a very upset winemaker Albane. Through her own professional integrity, Albane knew that she personally had to call and explain the (very) bad news. Our precious Cabernet Franc 2011 wine had turned to vinegar.

Albane is our organic viticultural leader and winemaker. She had designed and led the Cabernet Franc recovery programme in our Hourcat Sud vineyard. She had also trained her mother, Pascale, to become one of Bordeaux's most passionate caring "hands on" vigneronnes. This was an unhappy moment for both her and the project. Paradise Rescued is a close team.

Let's put this into perspective. In 2011, we had one block (Hourcat Sud) of Cabernet Franc vines producing one vat of beautiful wine. Our 2010 vintage had gone well, produced a wine significantly better than we had ever dreamed of and we were starting to see a strong market potential ahead. There are only a handful of 100% varietal Cabernet Franc producers in Bordeaux. We were opening our eyes to the possibility of a real niche brand opportunity. However, at a business and financial level no wine meant that there would be no income stream from that complete year of operation. All the vineyard's costs, wages and overheads would have to be simply stacked into the debt column with the hope of the previous year and the following year being able to cover them. In short, it was a potential business-terminating event.

I have reflected several times since on that phone call and Albane's bravery and strength of character in making it. I have also reflected on how I handled it. As a chemical engineer and senior manager, I was experienced in handling bad news and unexpected business events. In many instances around the world, the 'normal' management reaction is anger, aggressive disbelief and immediate blame transfer to the messenger brave enough to announce the "bad news". I have come to learn that this behaviour never actually changes the history of the event but significantly impacts on future communications and willingness to address a permanent solution. I am glad that I didn't make that same mistake, despite the gravity of the news.

Albane and I discussed the issue in our vat and she provided several potential solutions or opportunities to manage the problem. Our vat had been attacked by acetyl bacteria. The 2011 vintage had started early, continued dry through spring until finally some rain arrived in late June leading to an early harvest – our earliest harvest so far – in the third week of September. Very warm conditions prevailed through the fermentation and afterwards rendering the vintage (in Bordeaux terms) very vulnerable to this kind of event. All agricultural and viticultural products are fragile as nature is generally uncontrollable, unpredictable and at best only partially managed by human intervention.

In my own initial design of the winery and desperate but passionate efforts to find suitable committed wine making advice, I had used my seemingly good chemical engineering experience in another field of that science and incorrectly applied it to oenology – the science and engineering of wine making. In short the vat we were using was way too big for the smaller yield of the 2011 vintage, for which my design was completely inappropriate. I hadn't yet learnt that our grape harvest yields could be so variable in quantity and that we needed a different winery and fermentation vat design to manage that variability.

We could have legally blended the spoiled wine away, sold it off cheaply and recovered some cash or used a number of unauthorised means to recover value. As Albane and I discussed the options, I was staring directly at our Vision statement on the wall above me.

It contains a series of values imposed on a picture of a bottle. One of those lines simply reads 'Quality and Excellence'. As Albane offered up more persuasive suggestions, my response was a consistent but calm emphatic "Non". The wine didn't meet our standards. Any attempt to work around the issue could impact our reputation and future market / brand perception. It had to go to the distillery to be converted into vinegar. And that was that.

It would be easy to read our story and conclude that we were pretty smart and had risk-planned ourselves so well that we could take apparent minor setbacks coolly in our stride. Or that we were completely clueless and just plain lucky to still be in business? The truth is probably somewhere in between. The beauty of this book is that we can get up close and personal with our readers and share a lot of those things that we did well and similarly point out where we made some fundamental mistakes. But importantly, where we made those big mistakes, we have learnt from them and moved forward.

I naturally don't wish failure upon any human being as quite often many worse situations can come from it. However, I also understand that the majority of small start-up businesses fail. I hope that by my writing these lines that you, the reader, can learn sufficiently and quickly enough to avoid many of the mistakes we made. If we change the odds of failure by just one or two percent, then indeed that will bring significant success.

I also appreciate that a huge number of small businesses will fail between the time that I write this first chapter and I complete the last one. The reasons are many and varied and our experience will not be the same journey for everyone. But if we can at least help a few more businesses avoid our errors, then this will be very positive result.

Better still, a few more dreams will come true and more lives will have been fulfilled.

"So David", I hear you ask. "2011 was a complete failure for Paradise Rescued, right? How could you have been so stupid to have not seen it coming?"

Does this sound like your boss talking, your wise caring sharing partner or even your own 'self talk'?

Whilst typical throughout our world, I don't believe this kind of behaviour is productive in any way, unless you are about to be run over by a train which you still haven't seen coming! In most cases, almost all of us can quickly see the error of our ways. And if we miss the subtlety, our bank manager, accountant or lawyer will be swiftly on hand to advise us of the impending situation.

2011 was the worst ever year for Paradise Rescued but the one where we learnt the most – about our business and ourselves. It's all about the way we looked at things. If we had even slightly considered it a major business-ending event, indeed Paradise Rescued might have fallen. Had we not done enough planning up front, we would not have had enough money and capital to get us through to another productive season and vintage – even if the two that followed were also very challenging. On the other hand, if I personally had planned better and gained more knowledge about wine making and better advice etc, then perhaps we would have avoided the technical errors we made. But then again we might not have profited from the event and learnt as much as we did about 'small vat' or 'micro-cuvé wine production.

The Value of Failing

As we build businesses or tackle any major goal, vision or event, it is important to realise that there will be unforeseen errors, unpleasant events, natural or human aided disasters or downright failures. Dress them up any way you like, call them what you want but they will happen! No one has accomplished anything significant on this planet without having had a good number of major setbacks from which they have had to pick themselves up, get back on track again and move ahead.

Whilst I don't want for anyone to experience failure to the point of having to stop trying to achieve your objectives, I do wish that you encounter some significant setbacks and obstacles that help you question your strategy and determination. Only through the adversity, the struggle and your strength of leadership can you really succeed. If we never did any work to achieve our goals and success arrived 'free of charge', there wouldn't be a word for success. Nor any stories worth listening to. Life would be so easy that we, as a race, probably wouldn't even exist and we would not know the depth of our human potential.

To many readers, our whole project and goals may appear quite mad? That's the beauty of a dream and the possibility of doing and achieving something so unique and valuable – in our eyes. The brilliance of youth is that you lack fear and sense. The wisdom of increasing maturity is that you can hopefully apply some gained experience throughout the journey, providing you don't then lose the courage to ever begin.

I was lucky during a personal 'lost period' of my life to have read a lot of inspirational leadership material which had centered me well for the journey ahead. I didn't realise it at the time. In "From Cabbage Patch to Cabernet Franc" I wrote about becoming very serious in 'finding myself'; that slow period of thought and resilience building had prepared me well for the Paradise Rescued project.

We knew that we would hit a number of problems – maybe not as big as this one – but big enough anyway, such that a lot of resilience would be called for. Naturally as the leader, if I crumbled and gave up at the first sign of trouble, that would have been a career-ending step for the project. I knew how much personal commitment a few amazing people had put in up until that stage. Letting them down would not have been acceptable. Letting myself down would have defined my life poorly. Our motto is "Failure is not an option"

A strong risk awareness from an early stage, gained through years of working in a very smart but risk adverse alternative industry, also played a valuable part. Whilst I will never tell you that we had a perfect financial plan, we certainly had strong direction and had considered the economic outcomes carefully. Despite the huge family strains of paying down the loans on the different parcels of land, the project was sound enough to go forward on the basis that we could always afford the wages for our vineyard team – paid if needs be from our own pockets. This is usually what is known as a Director's loan! And of course we also knew that the process of making a great wine is a slow one, taking maybe a few years to achieve reasonable drinking quality. So, although we never really adequately considered the administration and marketing costs, we did have a contingency plan. We had thought about the risks – the 'what-if' scenarios and how to mitigate them, should they occur.

Being Values Driven

As I write this chapter, I have just had a quick check of our Facebook page and in our newsfeed popped up a picture plus quote from an Australian Leadership author by the name of Pat Mesiti. I have read a couple of Pat's books and admired his straightforward, uncomplicated, and passionate writing. His quote on the post picture read: "Drive your values, drive your decision making!" Wow, I thought, that's the other half of what I wanted to write in my first chapter! Bingo!

The decision making process during Albane's sad call wasn't actually that hard. Honestly, despite the loss and the knowledge that as a team we would have some tough work ahead to rebuild financially and mentally, the decision between us was quick to make. And the reason for that "easy decision" was that we had decided on our values - the values that we wanted Paradise Rescued as a sustainable company to have and to be.

I will write later in more depth about the power of having a clear Vision of your business, organisation or goal such that it pulls you forward. On our Vision, we had taken the time and put in considerable time into thinking through, then writing down, the values and operational guidelines for the project. They are unchanged today – after 5 years of operation, three and a half as a French registered private company!

On our Vision picture, the values are written in the middle of a bottle image and even titled Values. The first one is Quality and Excellence. I have always been driven throughout my career about producing high quality products, focused on achieving the best results possible. There might possibly be a debate or question about the price of our product. But never it's quality.

As I talked with Albane, I was able to look up from my desk, set my eyes onto the A3 Vision picture above my head and there was my answer – Quality and Excellence. If we wanted to aspire to that value, there was no way we could even contemplate selling a slightly vinegary wine. The answer was "Non"! The answer was all there ready for me in our Values.

I was fortunate to be scheduled to travel through Europe on business just a couple of weeks later and able to amend the schedule to pass through Cardan and catch up with not only Albane and Pascale but also all of our neighbours and friends who had participated – voluntarily as always – in that harvest, and who shared in our disappointment. This was also a critical step in both acknowledging the gravity of the issue but re-affirming our commitment to continue. Our picture was bigger than what one setback could create. It had happened but a bigger future lay ahead.

I have been asked many times about what our leadership response to the French Vineyard Operations team had been. I sense the human curiosity as to how I had reacted and whether we had fired the team and hired a new one. This is a fascinating debate. At the end of the day, the only person who was truly eligible for dismissal or a reprimand was myself. Not the team! One error can be a learning opportunity; a repeat mistake is a strong warning sign that the previous lesson was not learnt and a third strike requires a probable change in team leadership.

Our team did not deliberately set out to lose that vintage. Their biggest "punishment" was to their own pride, self-esteem and reputation. I didn't have to do anything. Other than to re-affirm our confidence and belief in their skill and competence and their leader's belief in the Vision we were all pursuing. We were sad for a while but we moved forward together again. We all share a little bit of the scar that 2011 had left on us, but it wasn't a gaping open wound.

Our values and Vision statement also talks about our culture of learning, innovation and people. I didn't ever think about it at the time but (thankfully) I had responded and reacted in a manner that reinforced the values expressed in our Vision. Yes the loss of the vat of wine was a hard blow, but the way we responded as leaders and professionals has created our own internal culture of openness, learning and continual improvement. Our attention to detail, passion to never ever again repeat 2011 now bonds our small team together like 'Superglue'. Everyone instinctively double-checks each other at every step of the process, asking supportive questions, encouraging and ensuring that that event does not recur. The quality of every subsequent vintage has been a step higher as the result. The passion to never repeat that vintage is huge.

In hindsight it now seems like a fairly straightforward telephone call. An issue, a discussion and a calm decision. But it was another defining moment in our journey.

Over the winter, there were a number of phone calls to and fro as we sought to answer the question of what had gone wrong. The acetyl bacteria issue that had impacted our vat wasn't something new nor was it confined just to our winery. By sharing our experience and seeking answers, we concluded that we needed smaller vats. In the spring of the following year, Albane, Pascale and I sat down over a glass of wine in their new back garden, reviewed the options available for change and selected four new small vats to replace our single larger one.

The plan to bounce forward – not just move forward – was set. Our so called "micro-vat" or "micro-cuvé" strategy was set in motion. This new approach has given us whole new dimension to wine making in our Bordeaux village of Cardan, enabling us now to closely discern among different areas of the vineyard and hand select different vats for different final wines.

#LeaderTips

At each stage of the book, there are so many lessons that I want to share from our stories and journey. At the end of every chapter I have tried distill the story into a few key learning's under the heading of '#LeaderTips'. The reason I have called it '#LeaderTips' is to highlight the fact that, most often, we don't have to develop something revolutionary ourselves but, as leaders, we do have to do something. To take the lead and make a change. Frequently we can just use someone else's good idea and apply it to our own business. Proudly! Why re-invent the wheel?

I have also left a separate page after each section for you the leader reader to record your lessons learnt. And more importantly, what leadership action you personally are going to take in the next 30 days as a result of what you have read in each chapter.

So here goes with my tips for you from Chapter 1:-

#LeaderTips:

1. Failure is an event. Period! Not a life or business termination point. Failures, unforeseen errors and mistakes will happen. Understand that fact. Learn from them and use them to motivate you to take things to a higher, better level in future.

2. Be prepared for adverse events. In fact, prepare for them, prepare for what could go wrong and then develop your organisational culture to have a plan to deal with them when they arrive.

3. Be a leader not a spoilt child when bad events happen, knowing that part of the responsibility rests with you. Act appropriately to manage the future and prevent recurrence.

4. Values – get them established and understood up front. When you need to have them, it is too late to make them up on the spot. Write them down, look at them often, share and teach them. But above all, live by them.

My three key learnings are:-

1. _____
   _____

2. _____
   _____

3. _____
   _____

My next 30 day action item is:-

_____
_____

2

Merlot the Magician

At the winery house in Cardan, when you are sitting on the small westerly facing terrace outside the kitchen, facing the magnificent orange sun as it begins it's descent for yet another day and you have a glass of wine in your hand, then you know you are in a special place.

Countless visitors over the years have attested to this - so I know it's not just me. I can't explain why or what - you just have to feel it. And to feel it you obviously have to be there!

When we bought the property in late winter 1992, it wasn't quite the right time of year to soak up that feeling. In fact 1992 was a wet cold summer so despite spending some time at the house before suddenly exiting to Australia, we didn't really get to appreciate that beautiful feeling of being on the terrace.

In September of 1993 we were finally able to properly "be at home " in Cardan for a four week spell. Sharing the house in the beautiful South West of France for the first time with our families back then was when I first started to get an overwhelming feeling that, for whatever reason, this locality – the house, community and surrounding country – would play a significant role in my life. It certainly wasn't a logical engineering type of thought, but the feeling was there. And every time I returned to Cardan, it simply became stronger, re-enforcing the need to focus this new found energy into something positive.

Back in 1992, the Hourcat Centre block, as we now call it, was in its prime, growing a small one tenth of a hectare of Merlot (our existing and current 'Old Vine Block') and about three times as much Sémillon vines. It was a beautiful small block, running north south along the entire western boundary of our property and was effectively a green barrier shielding us from the rest of civilisation. It created a very special haven for all who dwelled there.

In From Cabbage Patch to Cabernet Franc, I've narrated the story of how we (luckily) came to acquire the land when our neighbours needed to repair their farmhouse. We were able to help the Dados family out by purchasing this gently sloping block. Prior to our purchase, the vines had been so called 'cut to death' and effectively abandoned during the 2010 season. As we signed the compromis (purchase contract) in April 2011, we included a clause that allowed Pascale and the Paradise Rescued team to immediately take over the viticulture and management of the block.

We had a very specific strategy, plan and goal.

The Merlot Big Picture

Sometimes an opportunity presents itself that fits snugly with your overall business or life goals, values and direction. If the synergy or 'fit' is good, then you owe it to yourself to rapidly develop or modify your strategies and plans to accommodate that opportunity. This was what happened with the Hourcat

Sud Cabernet Franc block. We had an overall plan as a family about where we wanted to be and what we wanted to do. When that block suddenly became a must-have purchase; the earlier energy and thinking was transformed into creating what is now Paradise Rescued.

Hourcat Centre was different in that we had effectively been debating and negotiating its purchase for the last fifteen years in order to try and position ourselves as being first in line to buy the vineyard if and when the opportunity arose. This patient diplomatic approach had paid off well. But at the same time, we were playing a waiting game; preparing a plan for what we wanted to bring into reality on that block. We had a clear strategy, which we then integrated into the main Paradise Rescued Vision and plans at the point of purchase.

The Merlot wine grape varietal is an excellent vine to grow on the so-called Bordeaux Right bank – that huge vineyard area that sits to the north east of the River Garonne and includes all the sub regions of St.Emilion, Pomerol, Fronsac, Blaye etc, as well as our own small estate at Cardan. Merlot can be fickle but it generally thrives in the climate and clay/limestone soil found across that area. It is by far the most commonly grown wine varietal in Bordeaux and has set a benchmark for the world in terms of top-class Merlot wines. In great vintages, when Bordeaux Merlot is on song, it is a world-beater.

I had always presumed that Merlot would be part of our destiny. When – at the last moment – we found out that our first vineyard block in Hourcat Sud was planted in Cabernet Franc, we were initially quite taken aback. Then I realised how we could turn it to our advantage and create something almost unique in the Bordeaux region.

The strategy for Hourcat Centre was (and still is) quite simply to create a high quality Merlot vineyard and naturally produce an amazing wine as the result. From the outset, the vision, strategy, plans and goals were all aligned to create this small high quality Merlot vineyard. It was all about starting with that end in mind.

Once we had got to the point of buying the land and setting up the contract to make the purchase, I wanted to move quickly in order to limit the damage to the old Merlot vines caused by the cutting them to death in 2010.

Within a week of having signed the compromis, Pascale started work on the then 55 year old Merlot vines to prune them to our standards and re-invigorate them back to full health. Had we not had a clear direction, they might not have been recoverable and we would never have been able to appreciate the amazing small quantities of Merlot wine that this one tenth of a hectare block now produces.

Sadly the casualty would be the Sémillon vines which were even less happy with their earlier treatment. We knew that this would have to be the case. By mid-2011 this part of the Hourcat Centre vineyard was completely devastated . Working on our full rejuvenation plan was the best option by far. The vines were dug up in the middle of 2012 and the land prepared slowly over a two year period for replanting in mid-2014.

Ironically some of the Sémillons got the last laugh. As there was no physical separation between the two sections of grape varieties in that vineyard and the rows simply run into one another, about half a dozen Sémillons survived the dig up and have started to flourish at the top end of the Old Merlot block. This has created a little story and humorous memorial to their less fortunate colleagues. Their tradition lives on!

With plenty of love and care, the Old Block Merlot recovered very well. 2012 was a challenging vintage in Bordeaux with significant mildew attack and damage in many parts. Our first Merlot harvest was a success – vinified in the redesigned micro cuvé based winery, proving the overall potential of the Hourcat Centre block. It also showed us one other thing. That 100% Merlot from Cardan would be excellent but maybe a touch overpowering and lacking the finesse that a small quantity of Cabernet Franc would bring to its nose and length when blended. A new tradition was beginning…

Commitment to the Long Term Vision

It's good to have strong strategies and to commit to a path. A clear direction is very empowering to a team and sends a clear signal to one's customers about what to expect and what's coming up. Endless vacillation and re-analysis of a strategy creates significant indecision, lack of momentum and internal cynicism. However, there does always need to be some flexibility in the detail and implementation to allow optimisation, adjustment and continual improvement. We need to set our goals in concrete and our plans in sand.

As Paradise Rescued progressed and our brand plus our Cloud9 / BlockOne 100% varietal Cabernet Franc wines started to become known, it was clear that if we wanted to blend some Cabernet Franc together with the Merlot that we were going to produce, this would impact on the market that we were just creating. We were already struggling to supply sufficient quantity for a full varietal boutique Cabernet Franc. In other words, this might be counterproductive for the brand.

The Hourcat Centre block also includes a small 50m2 section that lies directly to the north and behind the winery. As excessively wet weather in both 2013 and 2014 slowed our plans to re-condition the soil and replant, we finally made the decision to replant this small section in Cabernet Franc not Merlot. This will not only give us the potential to provide some Cabernet Franc to blend with the Merlot but also allows us to assess the different location and aspects in respect to any future developments of our much cherished Cabernet Franc varietal wines. A small well considered change will bring considerable additional benefits.

Developing a new vineyard is a (very) long-term investment. Our Vision, strategy and goals are all focused on producing good fruit and excellent red wine. We will not be rushing the new vines into production. The new vineyard is a fundamental long-term act of faith in our sustainable business for the future spanning possibly the next 50 plus years. Another generation of the Paradise Rescued business will most likely be the main benefactors of this investment as the vines achieve their full potential.

Good wine is made from great fruit. Great fruit is only possible when you have healthy vines growing in a healthy soil and ecosystem. It will be worth the investment for the future in getting the soil well prepared and those young vines successfully started.

It is intended that we will grow and harvest a limited quantity of fruit in season 2018, four years after planting. There is no point in trying to force the young vines excessively in order to gain one or two extra vintages. This would ultimately reduce the quality and long term potential of those vines, and would be completely counter-productive to the project and its sustainability goals.

Instant Success Formulas in Life and Business

How often do we see this principle being violated in life and business? Intuitively we all know it makes sense. It seems that we increasingly succumb to potential short-term gains rather than delayed gratification. We live more and more in a world of 'overnight-ism' where instant success and wealth is expected. The extreme position is that we want to set up companies today, reap the rewards tomorrow and then retire the day after as multi-millionaires and apparently relax ever after! Creating wealth, be it personally or as a corporation, takes time and a lot of focused effort.

Very few people end up as overnight millionaires, but many of those who do, are generally clueless as to how they achieved it. This often then leads to their losing their new wealth almost as quickly as they gained it.

Business is very similar. If you want to create something sustainable and have lasting value it requires a lot of concentrated thought, planning, personal development and hard work. Overnight success in business is not a realistic goal.

Occasionally it may happen but be very careful if you do believe that you have seen it; don't assume that it all happened with nonchalant ease and minimal effort. It most likely didn't! The media doesn't show the 18 hour days, 7 day weeks, 5am starts and three failed previous efforts. That's not saleable news!

In life and in business we all have a choice to make. I call it the 'Pay now, play later' principle. We can choose to either 'play now and pay later or pay now and play later'. It's a pretty simple principle really and it applies equally well to business as it does to individual people's wealth, personal development and our overall lives. You have to make some choices. And, by the way, not making a choice or choosing not to make a choice, is still a conscious decision.

Private vs Public Company Thinking

In larger public corporations, the focus is often driven by short-term returns and the personal external demands of shareholders. Their strategy is focused on developing good returns for their clients, who frequently have limited or little real interest other than personal financial gain in the well-being and success of that company. Killing the goose prematurely in order to obtain the golden egg does not achieve a sustainable outcome.

However, the need to encourage and reasonably reward corporate performance, effectiveness and human creativity can be very healthy - this is at the very heart of capitalist thinking. If taken to extremes and pursued without regard to a longer-term outcome, the results are very destructive both for wealth creation and also all individuals involved in that organisation. Balance is required between reasonable reward and long term success that will continue to reward everyone concerned.

Small private businesses have an advantage in this respect. The shareholders are often the directors and direct beneficiaries of the company's activity. The pressures of short-term financial gain versus long-term reward are usually more tangible and evident to the small number of people directly concerned. Both the goose and the golden egg are much clearer. And the chances are high that those people making the strategic decisions will be there to collect the results – both good and bad! There is rarely a golden handshake and new promotion as recognition for the blunders or stock market success. In a private company, the buck starts and stops with the decision maker or Director. This can significantly contribute to a more balanced approach to developing a company.

When we purchased the very first block, Hourcat Sud planted with Cabernet Franc vines, it was in direct response to a significant threat to our personal lifestyle. The heritage of our village and the sustainability of the rural lifestyle was about to be invaded by Bordeaux commuters all wanting their own patch of paradise, but potentially turning the village into a dormitory town. Immediate action was required and a longer term overall organisational strategy has emerged from that moment of decision.

The Hourcat Centre Merlot vineyard was on our wish list for a good period of time. As anyone who has had the good fortune to taste a small sample of the tiny quantity of Merlot wine that we produced from the old vines in 2012 can testify, this vineyard has great potential and is very special. Indeed, as a wine Merlot is a magician but its greatest and most enduring spell has been the ability to teach us to create a long-term goal and strategy to harvest its richness.

#LeaderTips

1. Build your goals in concrete; make your plans in sand. Decide where you are going, what you want to achieve. Hold those goals firm. Set a strategy, make a plan and get started. Be prepared to change the plan (not the goal) if the results aren't fully working.

2. Think long term. As they say, Rome wasn't built in a day; nor will your business or life goals! Stay focused on what you want.

My three key learnings are:-

1. _____
   _____

2. _____
   _____

3. _____
   _____

My next 30 day action item is:-

_____
_____

3

Hanging Tough

After 2011, comes 2012! It's great to hope and assume that in a nice story book there is a happy follow-on from a very difficult point in life or business. But Walt Disney didn't write our scripts!

As we discussed in Chapter 1, 2011 was the vintage disaster year that no wine business wants to remember but can never forget. There was a resolve among the team to make amends for that year with the 2012 season.

As the Old Saying Goes: 'Man plans, God laughs!'

Between Pascale and Albane there was a commitment to take it all to the next level of professional viticultural management. Although continual improvement is one of our values, they decided not simply to bounce back but to bounce forward.

The climate responded in equal measure with one of the most humid mildew-inducing spring seasons seen in many years.

I remember talking to Pascale just after she arrived into the vineyards at Cardan one morning having traversed a good number of kilometers from her home by car. She was shocked at the number of vineyards she had just passed where the foliage had turned grey; such was the intensity of the mildew attacks.

Our own 'micro-treatment' strategy has always been to spray-treat the vines very gently but more frequently than most of our neighbours, so as to significantly reduce the total copper and sulphur dosage during the year. We also time our treatment just ahead of wet weather when the vines are most prone to damaging attack. During one of my calls with Pascale, she personally walked every row of vines whilst I was on the phone, checking under the leaves for any tell-tale signs of mildew. I think there were two leaves in the whole vineyard and those two were very swiftly removed.

The preceding winter had also been atypical. The vines had not fully shut down in late autumn. In some vineyards (particularly where early pruning had already been carried out) the vines were starting to push out their new buds, when a wave of heavy -15°C frosts and enduring freezing weather hit. As warmer weather suddenly returned in March, bud-burst began for real but with no homogeneity. April then went cold and wet as the topsy-turvy season got into a bad pattern. Unsurprisingly, flowering and setting of the grapes was also uneven. Endless rain then suddenly turned to hot sunshine without any rain and the pace of the vintage slowed back to a crawl. The véraison (colour change) of the grapes was very uneven requiring further manual work from the vineyard team to remove bunches that had only a small chance of ripening sufficiently.

We decided to open up the soil a little in early September and put a bit of vigour back into the vines. Nature rewarded our efforts almost immediately with an extended period of rain, which helped swell the grapes again and maintain enough momentum going into harvest.

Finally on the first weekend of October, we cautiously harvested the Cabernet Franc and our very first tiny Merlot harvest from the Old Vine block. The quantities were small, significantly down on our 2010 benchmark harvest. This was partly due to the season and partly due to our rigorous vineyard management and early de-selection of any fruit that would not ripen.

Our change in winery vat size and strategy was extremely timely in managing this type of vintage harvest.

Harvest Day finished at 6pm on the Saturday and less than an hour later, it started to rain and didn't stop for two days. We had achieved a significant team win. And although the wine was not yet made, the harvest was clean dry and healthy, requiring little further triage (hand selection) before crushing and putting it into the vats. It was clearly less fragile than 2011 and with the re-configured winery, steady vinification, racking and maturation, a new success was on its way.

The relief was evident across the team but particularly with Pascale and Albane who had felt the pain and loss of 2011 most intently, and clearly wanted to prove that they, and Team Paradise Rescued, could deliver. And they had. The full story of 2012, its taste and power are still awaiting full appreciation behind the opening vintage of 2010. The small quantity of Cabernet Franc produced is nicely ripe with an excellent concentration of fruit and finesse.

In many ways, the biggest and best surprise of 2012 was the Merlot! We always recognised that bringing the old vines back from being nearly dead would be a risky and challenging labour of love. Merlot is quite different to Cabernet Franc and we had to learn fast how to harness its brilliance. We had no idea what to expect from these old boys and whether they would make it or not. Back then, Merlot was less easy for us to really see its vintage wine brilliance. It is only after two or more years of experimentation, blending and tasting that we are starting to appreciate the potential (which we believed in) of the Merlot and, by contrast, the very positive but different aspects of our Cabernet Franc.

Leadership From the Other Side of the World

What we also saw from 2012 season was some amazing human aspects and traits that can come from an empowered team. Everything rises and falls on leadership. Leadership is critical at all times. It is like a deposit bank account. You have to keep adding to it continually in small or larger amounts. It is also very easy to lose the complete deposit of trust through one poorly thought through decision or error.

Leading a team and managing a business from 18,000 km away is a fantastic challenge. There are only limited opportunities for face-to-face discussion, review and meetings. For us, this makes the process very focused requiring not only good thinking but very specific decision making and rigorous follow-through. When we are together in small groups, the time is very intense as the natural energy of the team bounces off one another. When I am in Cardan, we like to review everything together from each row of vines, the vineyards, the wines and winery layout. By the end of each period together, the team knows what is required in the coming period and where the risks to our business can be found and will be managed.

As the leader, it is my role to not only set up the direction and goals, re-enforce the vision and values, but also help the team to see their own contribution in achieving them. I believe my leadership task is to review their work and activity but never take away their responsibility to manage and lead their own areas. At the same time, I have to help and facilitate them to solve their challenges within those areas without taking away their power to act and make it all happen.

At the end of every visit to Cardan - however long or short - I have to be confident that they are sufficiently empowered to deliver everything in their areas of responsibility, but still confident enough to proceed without needing to refer everything back to me. From 18,000 km away, managerial control is impractical and if excessive, it will kill the innovation and idea creation that we seek to continually encourage.

Furthermore, we all have different skills and personalities. This is critical and a key learning for any business, no matter what its size.

Team Work

Thank goodness Pascale and I have very different personalities! She is a very detailed and analytical person which balances well my being an achievement focused, big picture visionary. I love our vines but I could not provide the same care and attention that Pascale brings on a daily basis. Similarly Albane is a double degree qualified Oenology and Viticultural Engineer and is very skilled in training people in her areas of expertise. If I had hired two clones of myself, there would be a team of three of us standing at the end of each row of vines, motivating and encouraging them, telling them how brilliant they were going to be, but missing all the tiny touches that have made an amazing 100% varietal Cabernet Franc out of a cabbage patch.

In all organisations, there is a huge risk if each team member has the same skills and personality type as the leader. The leader's style and personal approach will duplicate – whether you like it or not! If all of his / her subordinates have the same skills and personality types, then the organisation will become very narrow in its outlook. Its ability to innovate and seek out the next improvement will disappear.

We all have different personalities and varying skills. No one of us is perfect in every aspect of life. Recognise that and harness it to your advantage. Find people who are different to yourself, empower them, trust them and watch them do wonderful things.

The full success story of vintage 2012 is a combination of a number of human factors. The resilience from all members of the team was immense. When our values based commitment to organic micro-treatment was so strong and the depth of bad weather plus mildew attack so intense, twice weekly treatments and vineyard inspections became the standard. That is hard work for everyone. As the leader I felt it too but in our situation my only real contribution was encouragement, support and re-enforcement of the goals that the team chose.

The other role of the leader of course is to celebrate a successful outcome!

After we had cleared everything away on that harvest day in 2012, there were only about four or five of the key players remaining. As the sun set we were sitting around the kitchen terrace table that only a few hours earlier had been covered in coffee mugs and croissants prior to starting the harvest – our French community harvest team work exceptionally well on that breakfast diet!

The fruit was safely in the vat cold soaking and ready to start fermentation very soon. Behind us, in the northern section of Hourcat Centre was the devastation of what remained of the dead Sémillon grapes awaiting removal and clearing away. It had been a long day; tired bodies were enjoying the last half glass of wine, water or beer. It was quiet at the table and silent in Cardan. Drizzle started to fall. We looked at one another and looked up at the rain starting to fall from above. Suddenly the smiles broke out and then our faces began to beam. As a team, we realised that we had stood up to the challenge. And won!

We were back in the game!

Sadly the 2012 story for many of our colleagues in the region was less positive. And that was before 2013 hit....

#LeaderTips

1. Never give up! When you think you're done, you are only just getting started. Human beings have amazing resilience. With the right leadership, incredible results can be achieved from seemingly hopeless situations.

2. Be patient. Patience, resilience and persistence are key. It may not always happen in the time frame that you want. Stay committed and focused on the goal. Build progressively towards your objectives.

3. Set and maintain high standards. Build great skills and systems (to sustain and maintain those skills and standards) around your areas of excellence. No matter how tough the going gets; do not ever compromise your standards.

My three key learnings are:-

1. _____
   _____

2. _____
   _____

3. _____
   _____

My next 30 day action item is:-

_____
_____

4

Reality Shock

In any commercial organisation, the rubber meets the road when you have to get some money for the wonderful product you have created! If you don't find anyone who wants to give you the right money for what you are offering, you don't have a business.

As a trained practicing engineer and senior manager, I can't recall how many times I have heard crazy (production) comments like: "Why don't they just sell it – it looks good enough to me"... or some oblique version of the same theme. As if 'they' have some magic wand that can convert any old junk into extreme amounts of cash. Wandering through various wineries I hear many winemakers struggling with the same issue. Often they are the owners having difficulty finding a suitably priced market for their products.

When we started out - before we had a vision, strategies, plans and goals - our main aim was to sustain the section of vineyard in front of our house and prevent over-development of our village, that otherwise might ultimately turn our rural holiday home into a suburban villa. We had no idea of what we could achieve nor where we could take it!

In fact all the business calculations had been done on the basis that we could afford to run the vineyard on a low cost basis for the next 35 years, give all the wine away to our friends and still have some left over for fun.

Mad? Kind? Unrealistic?

All of the above actually! Well a little of each mixed in together. Sheer determination and a strong desire not to ever see that piece of land – the Hourcat Sud block – created into a housing estate, has caused some interesting things to happen. Including this book and the first one, Paradise Rescued - From Cabbage Patch to Cabernet Franc!

Thankfully, I had learnt the value of understanding that "Between Stimulus and Response, there is a space. In that space is our power to choose our response" In other words, to think! And find a solution.

This principle can be applied to almost everything, every day. It is a great leadership principle in business or in life generally. When faced with any confronting situation, we can either react - fight or flee – or through the use of thought, we can adapt and respond in a manner that can give us advantage, or simply ignore the potentially perceived disadvantage.

We get to choose the manner in which we respond to particular situations.

If someone mistakenly opens the cage door from the lion's den and a hungry, under fed, 200 kilo lion with a bad attitude looks like he is coming right at you, then please quit the 'thinking' step, just run! But if someone just wants to take a bite out of you to make themselves feel better because they have had a rotten day, it probably isn't worth a fight! When a challenge comes up – as they always will – we can choose our response. Don't fight, don't run away. Think and respond proactively!

For team Paradise Rescued, we had to respond. From day one we knew that we would never be a big company; the economies of scale would never be on our side. Our values helped us to choose an organic hands-on, manually intensive approach in order to give ourselves the best chance of making high quality premium quality wine. What we hadn't appreciated was just how good that wine might be. Our dream and vision needed to grow!

A Big Surprise in Our First Barrel

Spring came ridiculously early in 2011. For family reasons, we had to be in Cardan in April. It was warm. Bud-burst on the vines was well past and the first baby shoots were growing rapidly. We had run off our first (2010) Cabernet Franc wine into barrels at the end of the previous December and a first serious testing was required. Pascale gathered our local team of commune supporters, and in the beautiful sunshine we sampled each of the eight barrels and each of us gave our considered sommelier's opinion.

As we tasted the first barrel, things went quiet. Around the small group, everyone had a sufficiently good palate and local knowledge. There was a good bit of sniff, swirl, sip, spit and swallowing going on! When wine tasters come across something different or special they usually become quiet and serious. Rather than engage in dialogue about their surprise discovery, they usually stare silently into their half full glass as if to focus all their attention on what is siting inside it and changing their paradigm. Pascale as our Vigneronne and Winery Operations Leader is very humble. If something is bad, you know it right away. Her ownership and passion for the Paradise Rescued vineyard and winery is total. When it's all going well or better than our expectations, her humility takes over. She was still staring at her glass, nodding her head slowly with a certain contentment.

But still no one spoke! The surprise was universal to the group! Pascale announced that we would taste the second barrel. This would then prove whether the first barrel was our benchmark or whether there had been 'freak of nature' outcome.

Young Bordeaux Cabernet Franc wine is divine. It is much more forward and less tannic than its Merlot or Cabernet Sauvignon siblings. At a young age, the nose is very cassis or blackcurrant-like, with the body starting to gain a little complexity from the oak maturation. As always with Cabernet Franc, the length was already fine, zingy, spicy and seemingly endless in the mouth.

By barrel number four, the conversation amongst the group was starting to open up. There was now a confidence that indeed the first barrel had not been a lucky one; every barrel – despite the usual differences that you get from each oak barrel – were tasting similarly and starting to get plenty of ooohs and aaaahs! I still couldn't believe the results and was desperately trying to cross calibrate with a good local wine that I already knew.

So I went and opened another Bordeaux wine from the cellar! I will never make that mistake again in the middle of a Paradise Rescued 'degustation' (tasting) session. There was no comparison. I didn't need to be a trained sommelier or Master of Wine to reach my conclusion. It was instantaneous – our Cabernet Franc was indeed remarkable and no accident. And even more surprising was the fact that it was only the first vintage! My sip of the other wine had confirmed my suspicion, but ruined my palate and ability to adequately taste the remaining four barrels. Of course, I re-tasted a week later to convince myself that I hadn't been dreaming!

From many summers spent wandering round tasting French wines, we also knew that achieving a reasonable price that could support the costs of our operation would be difficult. I call it a 'label based pricing club'! If your vineyard does not sit within specific areas or communes, you can't write the names of those famous villages on your label and consequently the French wine world won't give you the same price, no matter what! Appealing to the French President or any number of ruling bodies wasn't going to change the club rules any time soon.

Thinking Outside The Barrel

If 'Failure wasn't an option', there would only be one other way to make it all work. Find an alternative – a response!

Living in Australia gave us a different insight into a more modern 'new wine world' where, in more general terms, price was much better correlated with taste, level of excellence and quality. Pricing outside of Europe has little regard for 2,000 years of history and Roman inspired foresight, or investment on seemingly privileged blocks of land and areas in an "Old World" wine land. If you make good wine, generally you can obtain good prices.

Worse still in the French and European model, the difference in label and piece of ground made a very significant price differential that easily persuaded us that exporting the wine to Australia would be the way to go.

Do Your Sums

Whether it is a job, community project or passion to preserve a small community's rural heritage, sustainability can only be assured if the sums add up. The bottom line is if you don't make enough money over a period of time to exceed the costs spent, then sustaining a business or any organisation where money is the currency for survival is not going to work. It's all well and good having well-meaning and carefully thought out intentions but if no one is going to pay for their benefit, it will not continue for very long regardless of where in the world you want to pursue that dream.

It may or may not be ugly but the sums have to add up!

Even in our early days, when we were solely focused on just providing enough ongoing cash from our pockets to keep it all solvent, we knew that we personally would have to fund any shortfall in operating expenses from our own incomes. Given the size of the actual land investment, the majority of which was in a very expensive house-building zone, we quickly worked out that additional cash flow would indeed be useful and that the project needed to move from a labour of love to a commercial enterprise.

The basic calculations showed us what we needed to do. Even allowing for export and transport costs, subtracting the high liquor (Wine Equalisation Tax) or WET and Goods Service Tax GST, it was very clear that selling into the Australian market would be far more commercially rewarding. And given our local physical presence in Australia (and our absence from France much of the time), the decision was not a hard one to make. I recognised that financial calculations should drive that decision, not the emotional assumptions.

References

Victor Frankl "Between stimulus and response there is a space. In that space is our power to choose our response." Man's Search for Meaning

#LeaderTips

1. In between stimulus and response there is enough space for thought! When faced with a significant challenge, put your brain to work to solve the puzzle. Don't simply respond with the first emotional action that comes into your head.

2. In almost all areas of our lives, we have choices that we can make. Use that power of choice. Choose wisely. Decide what you want to have happen.

3. Check out the business model in terms of dollars and cents. Put in down on paper and see that the sums do add up and that you will make money. Think through and prepare a full business plan.

My three key learnings are:-

1. _____
   _____

2. _____
   _____

3. _____
   _____

My next 30 day action item is:-

_____
_____

5

Marketing is More Than Just Selling

Our local research at both ends of the world showed that a reasonably good quality wine producer with a much larger acreage than us but infinitely more experience would be retailing their average to good bottle at say Euro 8. (As I comment later, a sizeable amount of lower quality Bordeaux wines sell for just a couple of euros only!) Through much benchmarking with samples safely stacked in travel bags going both ways across the world, we were able to work out that a similar bottle would attract a price of possibly A$45 dollars. Allowing for everything we wrote above, that was still a big difference and justified the risk of moving the wine to Melbourne.

My background as a chemical engineer operating in a world

of commodity products had left me a little skeptical of what marketing really was. I had only witnessed a narrow slice of marketing, mainly accompanied by (good) account management, analytical (almost clinical) pricing driven mainly by primary international oil prices with limited opportunities for branding and product innovation. Despite the indifference and frequent ignorance of my manufacturing colleagues, I had frequently had the opportunity to work right at the interface between manufacturing, product development and marketing.

As a tiny niche wine company owner there was no interface – it was all one group! The feedback and connection between customers and the vineyard plus winery team is rapid. Even then, we have witnessed some resistance to change and embrace the wishes of the clientele.

Old Traditions Die Hard

Wonderful passionate people who produce products and goods often believe that what they make is the best. The passion is good. The belief that, because they make it and think it is the best, therefore the world of customers will agree, isn't necessarily the case. If we want products and businesses to be sustainable, we have to produce things that other people will buy and pay for at an appropriate price.

Therein Lies the Need for Marketing

Thankfully, once we had got past the early unrealistic and 'charity' stage, we realised that if we were going to achieve a better than average price for our increasingly apparent drop of heaven, we would need to effectively market our business and product in a wine world that mostly focuses on cost, and make price the least good reason for people to buy our wines, or not!

I sat down with my team and worked out a plan that would get our name and product out there. Our goal was to both develop a market through small high quality respected retailers, in parallel with a good number of private clients.

The Retail Experience

The response to our strategy quickly became clear. With a high priced Australian dollar, wine exports were under pressure and European competitive imports were arriving rapidly. I wanted to position our wine in the top sales bracket but the retailers were awash with selections at that price point. They also struggled to keep margins sufficiently high as the large chain liquor stores were driving overall retail prices down.

We were challenged in three key ways:

1) We didn't have a name or brand that anyone knew, so people wouldn't be walking off the street to ask for the product by name.

2) Cabernet Franc is almost an unheard of wine grape varietal in Australia. It is a very different more subtle and complex taste from what many Australians would regard as a good $50+ wine, namely a fairly heavy full-bodied Shiraz or Cabernet Sauvignon.

Overall everyone who tasted the wine loved it.

3) Retailers were reluctant to pay a sufficiently high enough price that would allow them to move enough bottles and overcome the above pressures. Nor at the same time would their wholesale price offer return us sufficient margin to justify producing wine at that level of quality and exporting it to the other side of the world. Pricing, return and margin were therefore our third and main challenge.

On the other side of the coin was the private client market. Every time we presented and tasted the Cabernet Franc to knowledgeable wine lovers and passionate tasters, we were frequently rewarded with new sales and better still customers who wanted to follow our journey, progress and success with interest. A small army of ambassadors were starting to form and organise their own media on our behalf.

The way forward was clear. We had a focus for our marketing.

Get Help

In the middle of all of this process was an engineer struggling to understand how a wine market and marketing worked. As a business leader, I have one specific characteristic or personality, which is very different from many of my peers in both big and small business. I know that I can't and won't have all the knowledge required to solve every problem. When I don't have the answer, I am not the slightest bit embarrassed to go and seek out someone else to help me learn.

I do not know enough about viticulture and oenology to be able to do the vineyard work or make wine. And, of course, I am not often in France making the challenge somewhat harder! With an engineering and science background, I have enough 'feel' to evaluate whether our team decisions are right or wrong. I then usually delegate the responsibility for the work to those who do know better than me, whilst always ensuring that there is sufficient communication to make the team confident with what they are doing and giving myself reassurance that the business is functioning properly.

Similarly, on the marketing side we started with not much more than some general experience. We sought out help where we thought we needed it, following the learning we were given and gaining hands on experience as we went.

I recall back in 2011 having a Skype conversation about social media with marketing expert and author Dixie Carlton. The short one line summary of that frantic half hour conversation was: "Don't undersell the value of our product" and: "Build a brand". Those words of wisdom stopped us from selling our wine short and locking ourselves into the Australian retail market. I got the same message from Small Business Advisor and Kochie's Business Building TV star Linda Hailey.

Better things were awaiting for us around the corner as we learnt to market more.

Never assume that because you are the CEO, CFO, CIO and almost every other position holder of your small company that you will be good at it. You can't and won't. When you need good advice, get the best you can afford and follow it. Don't let your ego trick you into believing that just because you are great at one thing, that you will automatically be similarly good at something else.

Focus on what you are good at and where you 'bring in the magic'. Then get good help where you need it. Listen to the advice and follow it, checking frequently that you have learnt the real lesson and not just the message that you wanted to hear.

#LeaderTips:

1.　　Just because you have a great product, there is no guarantee that it will sell. Start marketing.

2.　　If you lack skill and experience in a particular overall area of your business, go seek good help and advice. Follow it – don't pretend that your passion will overcome all obstacles.

3.　　Develop both a production and marketing strategy and plan. Write it down and use it to drive your annual goals.

My three key learnings are:-

1. _____
   _____

2. _____
   _____

3. _____
   _____

My next 30 day action item is:-

_____
_____

6

Social Media Menace

There are many more advantages than disadvantages to being a new player in an old and deeply established market. By that I refer to both the global wine trade and the business of being in Bordeaux as a wine producer.

Wine is a fascinating product. Although all wines have many similarities, they are also unique to a certain extent. Every wine is grown on different vines on different land with different topography, various micro-climates and very different viticultural and vinicultural practices.

Much like the people behind each winery!

I enjoyed touring the Bordeaux region for about a decade before starting Paradise Rescued, and in that time got to know a number of the better producers and owners who have made really good wine in their class – outside of the Grand Cru clubs.

I spent a lot of pleasing time researching them and equally as much in tasting their wines and listening to what they did.

In overall terms though, it was hard to work out what, if any, marketing system was being used by thousands of producers across the Bordeaux region. Clearly the more successful owners I visited had been able to re-invest in upgraded winery equipment and sustain their ongoing quality performance and business. But from the media reports, decreasing areas under cultivation, plus all the local evidence of increased house building; it was very evident that overall – excluding the Grand Cru clubs who were in a different league altogether – the Bordeaux wine industry was not in a good place.

Although the worldwide perceived reputation that Bordeaux holds for producing outstanding red wines does not necessarily reflect every vineyard in that gorgeous region, a lot of very good, well made, high quality wine is produced. That in part had assisted my decision to export our own product.

Bordeaux Remains a Benchmark in World Wine

So why were so many wine makers trying to get out? It didn't make sense. Surely they should be able to attract buyers from almost everywhere? Yet as I looked around even more and I compared what I saw happening in Bordeaux to what I had seen back in Australia, a pattern was starting to emerge.

As you drive around the vineyards of Bordeaux, it is easy to feel overwhelmed by the concentration of vines, vineyards and wineries. Every road seems to have a dozen Chateau signboards at their intersections and each Chateau had a sign on their wall. But as I looked deeper, I got to understand that that was the extent of their marketing and outreach. Many owners are descendants of founders in family businesses that are maybe five or more generations deep; large numbers therefore do have well-established lists of existing customers. But the wine world is changing. France drinks less than half of what it did twenty years ago and the markets continue to shift significantly.

Whilst very good Bordeaux wine sells for a hundred euros or (a lot) more, this represents a small minority of producers but the majority of the media. Consequently the world of wine consumers is led to believe that Bordeaux is very expensive. There is a lot of excellent value for money wine selling for Euro 10 or a bit less. But a large portion is sold for less than a couple of euros per litre. This was clearly not a happy business playground.

The deeper I researched, the more stunned I became. Many businesses did not even have a web site! It seemed that for many producers, their only form of marketing and advertising was their Château sign hanging on their winery or house. How could one of the most talked about products on the planet in one of the world's premier wine regions not have a visible platform to the world to promote their products? This revelation was a both a shock to me and a source of inspiration as it provided an opportunity to do something well that other Bordeaux wineries hadn't – it was an opportunity to stand out from a very big Bordeaux crowd.

However, we could barely afford a vigneronne and winemaker let alone a marketing campaign! So our ability to compete effectively against large wineries with expansive marketing budgets was a serious challenge.

Welcome to Social Media (SM) Marketing!

By the start of 2010, we were tentatively making decisions to get something happening. The environment around the house in Cardan went from holiday home to micro business HQ. All sorts of interesting people came and went as we started to get a wine company into motion.

Our daughter Lauren was there to join the ski holiday and take time away from her dance studies. Her boyfriend at that time (who was also called David and was therefore quickly dubbed David2) was also visiting, and we enjoyed quite a few visits from our neighbour Nathalie and her daughter Faustine. Many fun multilingual evenings were spent round the fireplace there.

The subject of the Hourcat Sud Cabernet Franc block vineyard was most often top of the discussion agenda.

Everybody seemed pleased that at least we had found a wonderful (untried at that point but subsequently hugely successful) vigneronne and vineyard adviser, which was a very positive start. So we quickly passed from vineyard operations to the big question of how we would market the product. At that stage, we had no idea of the wine's future quality and potential.

Face-WHAT?

It was that time in history when the word Facebook was moving from a young person's mobile phone play toy, to mainstream business and lifestyle 'must-have' product. The first wave of iPhones (Smartphones) and the ability to have internet access in the palm of one's hand, was heralding in a new digital age revolution. It would be fair to say that along with many others of my age and generation, I could easily have been classified back then as a social media dinosaur, doomed to be bypassed by the digital age as it roared on in.

It became clear in many of the conversations in front of the lounge room fire that Lauren, David2 and Nathalie were communicating a different jargon from the elder participants and seemed to be inter-communicating amongst themselves and others through this new media during mini-bouts of head down focused attention on their screens.

From almost the first moment that the project became a bit more than a dream, I had started taking lots of photographs; largely to help me understand the dimensions of the land during the negotiation and legal processes. I had observed that all of our friends who saw the photos were totally entranced by them and continued to ask for updates whenever we met up. The natural beauty of our small valley, house and vineyards always seemed to be something that we would want to use in our business and naturally on a web site when we eventually created one. What I hadn't (yet) seen was the power of social media and specifically photo and video visual media as part of it.

The junior team in the lounge room formed an ad hoc management committee - without my delegated authority - and voted 3-0 that I should set up a Paradise Rescued Facebook page! I abstained and my wife Maggie carried on reading her book. The motion was carried and the action delegated upwards to the Director!

The Facebook page was quickly created – by the Director – and the three voting members of the committee were rewarded by their appointments as Page administrators. Nathalie took up a couple of other local voluntary roles including local vineyard photographer. And our brand culture of visual media and presentation was created. David2 took the position of chief social media advisor. Lauren admired with amusement and lots of likes!

By year-end 2010, we had a full web site up and running which integrated our blog. I have written a short blog post every week since early 2010 - initially as a means of communicating our direction and intentions to our community – and increasingly as a marketing tool to showcase our brand and open style of approach. My strong passion was to create an enterprise that was sustainable, and involved the neighbours and members of the Cardan community, who had helped to extend these opportunities to us and were already participating in our business dreams.

Our website quickly became the Paradise Rescued 'go to place' central hub for our Facebook and Twitter accounts. Through our network within the social media networks, we made contact with leading social media expert and talented trainer, Zoe Wyatt, from "Social Media Shortcut". In a short space of time she took me from being a dinosaur to a modern day social media leader in the wine industry. The outreach and communications media that she helped to create, has enabled us to find previously unthinkable international contacts that has helped propel our wine from a nice private Bordeaux garage wine to a unique brand and much sought after Cabernet Franc varietal.

Better still Zoe loves her red wine and together with her husband Mark, they have gone on to become passionate Paradise Rescued ambassadors.

A quick note to you the reader - at this point I run the risk of your losing the thread of our story as I plunge into the world of Social Media. With a small number of exceptions in the Bordeaux wine world and fairly generally around the world I estimate, wine producers haven't (yet?) appreciated the social media advantage.

So maybe I should explain WHY we have successfully embraced and benefitted so well from social media. And then HOW we did it so that you too can benefit from the lessons we learned along the way.

#LeaderTips

1. Social media marketing is a very effective digital age marketing media open to the whole world on their devices through the push of a button on yours. Take it seriously and start using it to your advantage.

2. Put your ego in your pocket, get your phone out. Just because a school kid can do it, means you should be just as easily able to master it. He / she could be part of your next generation of customers.

3. Social media marketing offers an almost unique means of reaching everyone in the world in one step or click.

My three key learnings are:-

1. _____
   _____

2. _____
   _____

3. _____
   _____

My next 30 day action item is:-

_____
_____

7

The Power of
Social Media

Although we only have a small quantity of wine to sell from each vintage, finding, developing and keeping valued customers is a priority and not easy to achieve on a tiny budget. But I knew that if we were to charge a premium price for what is a very good, almost unique wine in Bordeaux, I had to find a way to identify who those customers were and give them a lot of value for their money.

Alternatively, I could have gone with the standard prices for wine from that part of the Bordeaux region. And you wouldn't be reading this book today!

The challenge for Paradise Rescued was to find a cost effective method of getting our name out to the world and making it stand out from the other tens of thousands of Bordeaux red wines. Then we had to persuade those customers that we did find, that ours was more than just another lovely well-made

French wine. Bordeaux is a massive wine producing region. Hundreds of square miles of vines go on non-stop from every direction from our house!!

Social media offered us that low cost route to market. A Facebook page, a Twitter account and a number of other media come for free. You can add on some helpful tools for a few dollars here and there, but essentially they are free marketing tools. This meant that we could compete on an almost level playing field along with the biggest and best Grand Cru Club player. On a social media level today, very few of the top-flight wineries participate seriously in social media. For the moment, they don't have to. They have the advantage of having established significant brands and alternative media presence over a number of decades that continue to maintain their prices at significant premia to the competition.

Outside of the Grand Cru's – and other exceptional wineries around the world – the business of being in the wine industry is quite different. The media don't simply walk up to your door seeking interviews, tastings and photo shoots. Then walk away happy to publish your media and marketing message out around the world. In Cardan, Bordeaux (and many other similar villages around France and the world), life is equally different. We have to create our own media, marketing messages and work out how to get them out there. All on our own! And this is the same for many other industries and businesses too.

If you are not marketing your product, who the hell is? And if you want to, how can you do it affordably and effectively?

In this rich digital information age two things are happening.

1) Not only do companies now have that ability to publish their own media out there but,

2) Customers have greater power to find exactly what they want. If they like it – or not – they have free access to share their views of the world in public with almost limited chance of the product or service provider being able to do anything about it!

And if they need something else, they simply 'Google' up an alternative supply?

Many customers today are no longer happy to accept that a wine with a classy label, price and pedigree is the best deal that they have bought from the local bottle shop. Instead they are keen to learn where it is from, how it is made and actively participate in its daily activity! As wines often pass through several hands en-route to their final consumers, the communication, ownership and passion for the product evaporates.

Social media provides a unique opportunity for distant customers and the source of origin producers to connect. Companies who do engage with their customers, get their feedback and re-design their services and products based on that feedback and are often those who have the best chance of ongoing sustainability.

Now producers are more easily able to locate customers via social media, and any organisation is able to share its values and stories with the segment of the market wanting those products.

Building a brand no longer depends on the size of your budget. You can now do it effectively without emptying the cash account.

How to Build a Brand Using Social Media

There is a general assumption out there in the world that because we can all access social media platforms for no more than the cost of your mobile phone and an internet access account, creating a suitable social media brand will be a walk in the park. Similarly it is also assumed that it should be equally simple to create a significant presence that the world will instantly love. If you listen to the average player on the street you might expect that everyone could create overnight brand successes from only a few hours work. And as everyone can have a go – with an almost zero entry barrier, starting from 10 years of age upwards – it is easy to believe that fame and fortune are only two steps further away. All bad incorrect assumptions!

Of course the world doesn't work like that and even several serious morning's work, no matter how good, will do little more than get you started. Social media marketing is a very serious marketing business, despite its low entry threshold.

Achieving a reasonable market position – just as with any traditional and more expensive marketing options – takes time and consistent persistent effort. And the right approach!

I have learnt, discussed and published a few steps, which have significantly aided our social media success in the wine world. I also believe these can equally well apply to most businesses from almost any industry. So here they are:-

1. Develop a Strategy.

This item is number 1 because it is THE most important one. I see so many small (wine and other) companies using social media as an extension of their marketing work without a specific plan, getting frustrated with their results and then abandoning their efforts and media in full view of the world.

I cannot stress enough the importance of knowing what you want your business/market to look like and where/how social media will help you achieve that. This will then help you decide:-

- how much time and how often you are willing to invest
- what social media tools you are going to use
- how you are going to use each tool
- which of your marketing objectives will benefit from applying social media?

2. Get Professional help and advice

It seems intelligent logic that because so many people on our planet are using some sort of Social Media that you can become part of the excited crowd and you will have happy customers for life! WRONG!!

Most social media are well adapted for general social communication, sharing information, media etc. Making it work for your specific business is very different. You want to stand out from the crowd, not be part of it! To make this happen and help you develop and use the available tools to achieve the best results, you will need professional help and training.

Wine producers: when you first started – did you just plant some vines and start fermenting in the first vessel that came to hand? I doubt it (and I hope not!) – you sought help and guidance from people with a lot more experience than you had. Social Media Marketing is the same.

3. More dialogue, less broadcast and target, target, target!

Because much of social media looks like a world of people just broadcasting media to anyone and everyone who might want to listen, it is an error to think that this is the model one should follow. Yes, there is a need to promote one's material to a wide audience of people in order to attract new customers and partners. But the unique feature of Social Media is that it gives us not only the opportunity to find new friends and customers, but also to talk directly to our existing customers in a manner that has never been available before.

We can choose to send different media to specific audiences depending on what they want and when they want to hear it! Even better still, we can open up a discussion, obtain service feedback and learn how to improve their experience of our product.

In short, think of ways to have discussions with your current and future clients. Ask them what they think!

4. Choose Your Tools and Focus Your Attention

Every week it seems there is a new social media product or tool available. Don't try to use everything! Once you have your plan organised, decide which tools are going to give you the best leverage of your time. Also work out which media your customers are already using, then go and join them there.

Once you have some engagement and dialogue, stay there and maximise the use of those specific media options. Also find and use tools such as Hootsuite that give you a good overview of the different media you are using and allow you to optimise your time involvement.

5. Study the feedback

Another big advantage of the Social Media world is that you can always see what is going on. You can get feedback on each post and everything you do. Study your Google analytics, Facebook Insights, etc and learn what your customers liked and didn't like. Adjust your tactical plans to suit what your market wants. If they like more pictures, give them pictures and build your brand. If they like a full and open dialogue on your blog about your winemaking, talk to them about that.

And if something isn't working, change it. Try new initiatives. Ask your audience if they want something new. And work out what success looks like.

6. Let's get visual

"A picture paints a thousand words" runs the old adage. In social media the multiplier could be many of millions. The use of images is the main trigger for the success of social media. Without it, emails and the telephone would still be in charge.

Human beings function through five senses:-
1. Sound, 2. Sight, 3. Smell, 4. Feeling, and 5. Taste

In the world before social media, sound and sight of the written word (as in letters, emails and books) were the controlling senses. With the arrival of social media, our communication power has been dramatically extended. Now we can use both still and moving images. The success of YouTube, Instagram, Pinterest and all new media (to greater or lesser extent) stems from the use and desire to share using a second key human sense: Sight.

The integration and use of visual media across all your platforms is a must. Take photos, write about them, use keywords to position them and help online searchers to find them.

7. Do not expect overnight success

Building a brand or product position in a market is a marathon not a sprint. Your traditional marketing required consistent steady effort. Social media is the same, only different in its approach and ability to more closely engage with customers.

I see so many people start enthusiastically, collect the 'likes' of their personal friends and then when the flow of support stops, so do they. Push on – that's when you really learn about your market and customers' wants, needs, frustrations, and desires.

Remember that internet media is very, very public! So if you start something but don't continue, what you started (and stopped!) is still there and visible to all. An abandoned or poorly used social media tool is still in full public view even if you are not looking at it. This can send an even worse message about your brand and products.

Never believe that Social Media is a 'silver bullet' that will fix your sales and marketing for ever. It won't! But with a well thought out plan and consistent sustained effort, it will create a lot of good opportunities and a loyal customer base that can lead you to a much stronger overall business. Social Media is a very sound platform for building a (micro) brand!

If you don't see immediate positive results, continue steadily building your media. Your business most likely wasn't created overnight and your new social media marketing program won't be either. Persist with strategy!

#LeaderTips

1. The key to social media marketing is less about finding customers and contacts but more about keeping them and helping them to become your best advocates.

2. Never forget that as social media marketers, our role is to host and take part in a conversation. Don't broadcast! Engage with your customers and fans.

My three key learnings are:-

1. _____
   _____

2. _____
   _____

3. _____
   _____

My next 30 day action item is:-

_____
_____

8

Brand on the run

You either get it, or you don't. It's that simple!

As a manufacturing plant and site manager with experience in many parts of the world, I didn't get it. It wasn't even part of the vocabulary or jargon. In fact whenever possible, I went out of way to eliminate it - I kept it off the plant at almost all cost. When the head office 'marketing suits' turned up and they couldn't find any other reason to convince us to make the next wonder grade, out came the 'brand' word! If there was no compelling business logic, then 'brand spin' had to be added in.

Thankfully in life, all of us can change if we want to. And if you do allow another point of view into your thinking space, paradigms can be altered and incredible changes can occur.

I certainly didn't get it, but I do now. I can almost hear my engineering colleagues groan with dismay. Many of them will never get it! They simply fall prey to the magic every day even though they won't ever admit or consistently deny it. It's like gravity – whether you believe it or not – it still applies.

The apple always falls out of the tree! It's the physics of marketing – it's called a brand.

What is a Brand Anyway?

Dixie Carlton, whom I referred to earlier, said to me 'Build a Brand'. But what exactly is a brand? Her definition is this: It's the intangible 'everything' about the business of what you do and how that is communicated to your market.

In many ways that is the trick of a brand. We can't define exactly what it means to us but we know what we want to have. We use and buy it for whatever function or service we associate it with. It is significantly built on individual perception of value. It goes to the heart of what we want and are maybe prepared to pay. It's a different price for something which we see as having higher or lower value. It can overcome the mental processes of normal analytical logic.

Carlton also maintains that if you have a strong brand, your marketing works a lot harder for you to ensure all your advertising and media is more effective.

Which is why in the wine business, brand is so important. We produce wine from a region, which all falls under the general catchall name of Bordeaux Wine, comprising more 8,500 producers and probably five or more times that number of total wine labels. Branding is our significant way of standing out from the ranks of a massive wine crowd. Without developing awareness of our company through branding, our tiny size and likely inability to be heard above the noise of Bordeaux, would have ultimately consigned us to an early bankruptcy.

Creating a brand is not an instant process. We cannot create an overnight brand, even with a heap of money and advertising. It takes a while for that product or service to gain momentum, acceptance and the right value in the mind of any consumer.

A brand creates perception and expectations about quality, price, functionality, service, durability, pleasure and so on. Through the history, daily action and media we come to build a mental but non-analytical picture of what a particular brand can deliver for us. Quite often, it transcends logic and left-brain thinking.

Which is why it is so hard to accurately define what a brand is!

For us as a wine business, branding is critical as there are so many different wines available. One way to help our wine and products stand out from the crowd is to brand them in a sufficiently strong and deliberate manner that allows potential customers to recognise them easily and be confident when buying them.

For us, quite simply our story – the story of Paradise Rescued – makes up the fundamental elements of our brand.

Empowered By a Logo

I wrote in From Cabbage Patch to Cabernet Franc about how the combination of some magic design work and the clear communication of our community established something unique for Paradise Rescued and a solid base from which to build a brand.

When I look at our logo and how we have branded our products, I feel enormous pride. Then I see what our customers' reactions are, which further strengthens my feelings.

What Tricia Wiles at Sweet Graphic Design did so skillfully was to capture – almost project – the value and direction of our wine and business. The logo reflects, through its colour scheme, finesse of design and high quality - the values we have for our organisation, products, and intended future. It represents us and we represent it.

Out of Control... Really?

Anyone who knows me reasonably well, also knows that I proudly wear our brand as much and as often as I can get away with. We don't (yet!) have Paradise Rescued underwear nor do I sleep in one of our T-shirts, but whenever there is an opportunity to proudly display the tangible aspects of our brand, I will. I am the very proud and passionate owner of something that I believe is making a positive contribution to our community in France and hopefully bringing quality, pleasure, happiness and an inspiring story to many other people around the world.

As an engineer, I didn't get it. Now I constantly see value in it every second and in multiple ways. I want our team - the people who contribute to the creation and sustainability of our brand, to feel the same way and share in that successful identity. And through what we do and create, we want our customers to be our best brand ambassadors and be proudly sharing all parts of Paradise Rescued with their friends.

I can't understand why any successful business that wants to use their good reputation to gain and maintain customers doesn't even have its employees wearing their logo. I visit many wineries and businesses where team members wear the name of someone else's brand on their shirts! Why would you provide free advertising for someone else's business, assuming that they aren't paying you for it?

We take the view that if it is ours and it is part of our business – be it letterhead, email signature address, website, equipment, documents, computer, iPad, barrels, vats, merchandise, tools, anything – it should be branded with our beautiful logo and/or website address. I am sure that there a few exceptions lurking somewhere in our organisation that I have yet to get to, but generally where we want to be known and clearly identified, it shall be branded. End of story.

Change Your Lens – Experience Your Business as a Brand

Once I started to see our business as a brand (supported by specifically identified brand values) and became confident to talk about it in that fashion, the Paradise Rescued brand became the lens through which we started to consider our decisions.

Let me explain – I can envision some confused readers at this key point!

As I will expand on later, our organisation has a very clear Vision and Mission. We all know what we are doing and why we are doing it. Achieving results can sometimes be easy, but with unpleasant methods. I wanted a great outcome for Paradise Rescued, and for the way in which we did it to be effective so that it took the community we serve, our customers, fans, and our whole team with me on the journey.

Don't get me wrong, there have been many unpleasant decisions along the way. This has been more like traversing a theme park than a ride down a country lane! Those decisions had to always take our values into account because ultimately our values are at the core of how we wish our company to be seen as a brand. Therefore every time we decide on a change or new direction, we ask the question: "What does that change do for our brand? How will this impact on our customer's perception and expectations of us and our brand?"

We use our brand and values as the lens through which we focus our decision making. And that applies to everything from a change in service, product or investment.

Protect Your Brand, Protect Your IP

As a business owner, I am very conscious of the need to protect our intellectual property (IP). For a high tech business this may mean your technology. In most businesses, operating in a competitive world, it means how do you sustain that lead or 'edge' over the competition? How do you maintain and protect that special something which makes you and your brand stand out from the crowd? And how do you defend it?

As wine producers, we have applied some very basic, but detailed, organic viticulture practices and followed through with a similar semi-traditional combination of winemaking skills. We use technology where it makes good sense in producing a great product but not if it violates basic principles and simply creates a short cut. In overall technical terms, we don't do anything specifically different that one of our so-called competitors couldn't also do. The industry is sufficiently large and open that if we lack any particular skills or knowledge, we can easily go and find the skills to fill our deficiency in knowledge or equipment. We then adapt those human skills to get what we think is the best from our vineyard, continuing at every stage to improve the health of the soil, vines and vineyard ecosystems. It's a continual process of further fine tuning our winemaking skills to adapt to each particular vintage.

So we really can't differentiate ourselves to our potential audience of customers in the viticultural or oenological areas. In the marketing area, we can focus our attention on a particular market segment and locality – as we do. We can and do leverage the use of social media, which others could and may replicate. The only area where we create a difference is by our branding and how we use it.

Brand Paradise Rescued is its own IP! And that is something that we can both protect and then enhance as well as it being the means by which we differentiate ourselves from other wine companies.

To the right of where I am writing this chapter, I have a small dossier which contains the certificate recording the registration of our Trade Mark for the Paradise Rescued logo within the class of business where we operate.

Hands off our IP! But please protect yours.

#LeaderTips

1. Take some time to evaluate what is the real IP of your business or organisation. What would you least like to have stolen from it? Discover your 'magic'.

2. Consider carefully how to protect and safeguard it from theft, copying or misuse. Guard it with your life!

My three key learnings are:-

1. _____
   _____

2. _____
   _____

3. _____
   _____

My next 30 day action item is:-

_____
_____

9

Good Morning New York

The Tweet That Unlocked a New Story

I was doing my nightly scan of the social media, adding a little new material and checking on the progress of our different platforms. As usual on Twitter, there's always plenty happening so I checked the notifications. In amongst it all was a direct mail (DM).

I opened the DM and read the message: "You've got reach I've got reach Let's talk and make $$$$$$$$$"

That wasn't a standard Australian business greeting for sure! The author was Monika Elling, CEO and Founder of "Foundations Marketing Group", New York Luxury Brands Communications and Marketing Consultants.

When you have been out there trying to create some noise to be heard, then these kind of messages don't come very often.

One is more used to hearing a crescendo of "No, no and more no"! I can't remember what I replied – I was still stunned by the incoming DM. But the essence of my reply would have been 'Absolutely'. And naturally that's what happened.

I also remember the resulting first phone call. It wasn't easy to connect because the timing of our first telephone call tragically coincided with the Boston marathon bombing. But a start had been made. Monika had seen all of our work on the social and web media, read all about our project and was interested in tasting our Cabernet Franc. She had a personal liking for Cabernet Franc as a wine varietal, a new visionary view of what marketing wine would look like in the future and the strong positive contribution that social media marketing would bring to our industry.

Challenge number one was learning how to move bottles around the world and comply with different (often conflicting) international regulations. Some five weeks later, I was speaking /chairing a double venue conference and as the organisers already had a keen interest in our micro brand wine business, I was fortunate to be introduced to one of the key note speakers, Jeff Dudley, whose second career was in reliability engineering consulting and setting up a small vineyard in Michigan. He was fascinated by our project and was keen to not only to taste our wine but also to buy some bottles as gifts for the organisers.

The second week of the conference was in Perth and the trusty Paradise Rescued wine case went too, full to capacity. Jeff also bought some bottles to take back with him to the USA as well as one bottle to kindly post on to New York for us. This was a very elegant solution to the expensive process of airmailing a bottle from Australia. The wine case went back empty to Melbourne - a good week out of the office!

Although it all took its time to happen, the single bottle of Cloud9 Cabernet Franc made it successfully to New York. From the squeals clearly audible on social media, with pictures attached, I think the package must have remained intact for no more than about ten minutes after its arrival in the Foundations

Marketing Group office! The result was a thumbs up from someone who knows her international wines and Cabernet Franc well. A promising start – time to think a bit bigger!

Could We Dream Bigger Than a Small Village?

As you can see from earlier chapters, although our overall vision to sustain our small section of village and vineyard was pretty damn ambitious, it had not even contemplated the idea of becoming a niche micro brand wine company selling into New York and surrounding states, which is arguably one of the holy grails for luxury brands in the world.

Monika was about to change all that!

Our next phone call was different. The tonality had quickly moved from 'wouldn't it be nice' to 'how soon can we make it happen?' and 'how much?'. Monika was keen to understand how much we could produce and how fast we could get it to New York. She was clearly taken aback by our (tiny) size and the fact that the wine was already in Australia. She wanted to import 1,000 bottles per month at a nice price. We suddenly wished we had a whole lot more wine and had made a successful vintage in 2011 as well.

You can't alter the current state; you can only change the future. And in the wine business, planting or buying vineyards takes (a lot of) time.

Up until this point in time, we had come to believe that Cabernet Franc was the poor cousin in the 'Big Bordeaux Trilogy' of Merlot, Cabernet Sauvignon and Cabernet Franc that makes up almost all Bordeaux red wine. Whilst we thought that this sentiment was unfair, the rate of pulling up our much loved varietal around the vineyards of Bordeaux and seeming lack of mention or respect for its characteristics and contribution to the blended wines was telling the story in the market place. And as a varietal in Australia, it barely rates a mention.

We learnt an important lesson. Paradise Rescued could be a great niche brand and this naturally fitted with our size as a vineyard and as a producer. And, if branded well, niche can command a different and unique respect. Whilst Bordeaux was ripping up its Cabernet Franc vines, we started to set about making a name for ourselves as a single varietal producer, putting our apparent disadvantages to good use. One of the world's smallest wineries suddenly had a big advantage.

We were starting to learn what the right marketing could do for us.

In October 2013, following a truly tough vintage and challenging harvest, Monika flew over from New York to Cardan to check us all out. She travelled overnight, arriving via Paris into Bordeaux during the early afternoon.

'Press day' – that's wine press day, not the media type of press day – is one of the toughest and messiest days on the winery calendar. It's generally not one of the best days to invite visitors or customers to one's winery. The young wine juice is run out of each vat into a clean vat and the solids which remain behind at the bottom are dug out and pressed to produce the so called press wine which can be partly or wholly re-incorporated back into the wine as the winemaker decides. There is a lot of manual work, a lot of red grape skins, plenty of cleaning water, several red stained tee-shirts and a small press in the middle of the winery forecourt.

Monika was warmly welcomed to the team, presented with her Paradise Rescued T-shirt and promptly expected to join the work team! After final wash up, there was a good sit down and discussion around the team workbench with a bottle of Cloud9 Cabernet Franc for re-tasting and a good bilingual discussion about the wine and the very difficult harvest that had just taken place. We had a newly respected team member and the start of a partnership was born.

The Roller Coaster of Success

It is easy to assume from the outside that in the world's largest economy, access to that market would be reasonably straightforward. I wish!

Monika worked with us every step of the way. And there were many steps! Following the end of prohibition in the USA in 1933, a three-tier system of alcohol was legislated requiring all liquor business to be divided into three sections namely producers (and importers), distributors and retailers. As a potential importing producer, one has to effectively find the right importer and then the following two steps in the system: to get one's product to the end customer. Not easy unless you have a good partner.

Thankfully we now had a good local partner to help us navigate this difficult process and establish sufficient influence plus brand awareness at the consumer level.

Working with Foundations Marketing and their nominated importer, the first step was to register our wine name and label and obtain a COLA (Certificate of Label Approval). There was already a winery in the USA called Cloud9 so our application was rejected. This meant that our lovely brand name – Cloud9 – could not be used on any bottle sold into the USA. Effectively, if we wanted to continue to sell our wine in the USA, we had to change the wine name, label and obligatory back information to meet the USA requirements.

Every time delay meant that the wine became that much older. Initially we were scared because we doubted our wine's ability to 'go the distance' and improve with age or at least maintain its current lovely quality. Happily – or should I say very happily – we had once again seriously underestimated almost every aspect of our first Cabernet Franc wine. At every tasting, it actually was getting better, not worse. It had both great quality and good longevity. And a great bright future ahead of it.

Leading in towards the end of the year, the team struggled to find a new name. The struggle was probably in part due to our unhappiness at having to throw off the Cloud9 label. We had become very used to using the "We're on Cloud9" tagline. It had well symbolised the courage and optimistic determination of our brand and story thus far.

B1ockOne was chosen as the new name for our Paradise Rescued Cabernet Franc.

Initially of course we took a while to adapt to the new name. Now a year or so further on, it seems more natural and is starting to happily corral a number of tag lines and marketing approaches. It did symbolise that there is only one block, its uniqueness and still continue our 'number plus name' tradition, which also works very nicely as a hashtag too! Tricia redesigned the label beautifully, taking it up another level to reflect its new quality and brand perception. The application went back in for COLA approval again and there it stayed and stayed for what seemed like an eternity. Finally one year after our initial application, COLA approval was given and we were ready to start the rebranding and export process.

Every bottle had to be double de-labelled, re-labelled and boxed into new cartons. Then finally, we could make up the first pallet load and initiate the export process.

Long live B1ockOne Cabernet Franc! One Block. One vineyard. One wine.

#LeaderTips

1. Good wine doesn't make a winery business. It certainly helps but it is only the start.

2. Think as big as you possibly can about the market outcome you would like. Develop and step up your marketing strategy to meet that objective.

3. There is a price to pay to make good wine. There is an equal price to pay to find the right market.

My three key learnings are:-

1. _____
   _____

2. _____
   _____

3. _____
   _____

My next 30 day action item is:-

_____
_____

10

Why Your Mission is Vital

Paradise Rescued started business with one big advantage that most businesses don't seem to have, but must seek out and develop. As this book progresses, you will see that we have added a couple more aces that drive our brand and business.

For Paradise Rescued, Mission – Vision – Passion has become a mantra, a tag line and the core of our business.

But let's start with Mission.

As I remarked previously in Cabbage Patch to Cabernet Franc, most organisations get completely confused between and about Mission and Vision. And which is which! A view, I might add, is similarly documented by a small number of other observers and writers. When I see a poor lonely Mission statement hanging in the foyer of a company's reception area, I often wonder how it came into being and then onto that wall, gathering dust and usually looking like a memorial to a distant poorly understood business school lecture.

If you have to go looking for a Mission (statement), you clearly don't know what it is and why you are in that place doing what you are doing. If you are the owner of the company and/or CEO and this is the question you are asking yourself right now, then this is sad. If you really have little idea why you are there, how do you intend to lead and inspire a team to achieve great results, products, goods, services or whatever?

Thinking back, I can't remember how many management groups or leadership teams I have sat in having the debate and how many times I have been howled down by more senior colleagues who substituted rank and ignorance in place of knowledge that they should have had at their level. They all should have done their homework before such discussions. But more importantly, they should have understood the criticality and meaning of Mission and why it should have been central to the organisation that they were part of leading.

Maybe it is such a confusing concept, that most organisations leave it alone. And to a large extent in public companies, everyone gets confused with shareholder value, profit and money in general and then they start to believe that that is the only reason they exist.

Financial results are an outcome not a reason to be there. If our reason for doing something – either as an individual or an organisation - is purely, simply and only based around how much money can be earned, then it simply won't be sustainable or leave any tangible legacy. Money is not a mission for doing something; it is a result of having successfully realised something of value.

The history of Paradise Rescued started with us wanting to do something very specific. We wanted to essentially protect the land around us, and our neighbours in the village of Cardan, from being overrun by a tsunami of housing development.

Having succeeded in buying the land – which happened to have vines on it in one of the world's greatest vineyards – we had to decide how to best use the land in order to sustain and maintain it in the future for the community.

Our Mission was born out of a definite need for ongoing action. The purpose was very clear. There was no need for a management 'away-day' to engrave a tablet of wisdom. From the first day of operation and subsequently thereafter every new participant, partner and supporter knows why we are there, doing what we do.

And everyone also knows that for us – because of our origin, beginnings and strong linkage into our community – failure can never be an option! That's the way we have to look at our business and project, otherwise we will let our community down. The picture postcard westerly view over the Hourcat Sud Cabernet Franc vineyard block towards the 12th century church on the hill beyond captures it perfectly.

These last few paragraphs have been written – from the heart - in such a way that the reader will have been able to see certain key words and themes that identify some core components of a Mission.

So what should a Mission contain? What's its purpose? And why bother at all? Too many questions too fast! Let me take them one at a time....

A Mission for an Organisation Should:

- define why that organisation exists,
- state what is its purpose and
- record how it achieves its purpose

The Mission should be succinctly written in just a couple of sentences and in such a manner that it can be easily remembered and recited back to anyone who asks.

It should be the answer to a marketer or sales manager's 'dream question' – "So what do you do?". "Thank you for asking. At Paradise Rescued, we…"

For me, the Mission answers the questions of WHY an organisation exists and HOW does it achieve it. Note I specifically haven't focussed on the word 'what' as I like to reserve that for later when we talk Vision!

Your Mission doesn't have to be brilliant. I am sure many of our readers here will want to take out their pens and 'improve' our Mission when they read it below. Thank you for the kind offer! But please put that energy into your own Mission in such a way that it succinctly defines the purpose of your organisation effectively.

The purpose of having a Mission is to provide that clarity for everyone in the organisation, customers, suppliers and stakeholders. It should create a clear understanding of the organisation's role and reason for existing. It should help everyone to better understand decisions and enhance – it provides a filter, funnel or simple check whether something is right or wrong and helps to engage the efforts of everyone around that group. It should answer the question "Is that our role?" or "Does that fit here?"

The final question is "Why bother at all?" And it is a good question, too!

It is always better to have a well-written Mission (and Vision) than not! It should be the means by which the Company understands and aligns its role and direction. If not done correctly, nor actively deployed and discussed within and outside the organisation, it merely becomes an unloved document in the reception area. And therefore it is probably creating more harm than good and not worth having.

A good Mission will contain one or more verbs, ie 'doing' type words. It often starts with something like "Our Mission is to..." And is immediately followed by an 'action' type verb.

And almost finally, you will also notice that I deliberately don't use the word 'statement' as in 'Mission Statement'. A Mission certainly does need to be written down, but above everything else it has to be communicated, taught and discussed inside an organisation. Which is why I wrote above that everyone in and around Paradise Rescued gets to understand, from the first moment that they begin to work with or alongside us, why we do what we do and how. We explain face to face why we do what we do and why we have so much passion. Everyone gets it in a heartbeat! We want them too to help us to make a difference! We want them to actively share their passion too.

Which of course brings us to our Paradise Rescued Mission:

"Our mission is to protect the heritage of the beautiful village of Cardan, France, sustain its rural community and to hand produce high quality organic wines."

#LeaderTips

1. Clearly evaluate, understand and communicate the purpose of your business or organisation. WHY you are doing your activity? That is then your Mission.

2. Once you have defined your Mission, use, deploy and communicate it as much as possible. Do not simply print it out and frame it on the wall. It must live in the organisation.

My three key learnings are:-

1. _____
_____

2. _____
_____

3. _____
_____

My next 30 day action item is:-

_____
_____

11

Vision is more than a Dream

I am going to open this chapter with two conflicting quotes. The first one is the traditional timeless discouraging putdown phrase which reads:

"I'll believe it when I see it."

Meaning that I won't actually believe what you have just told me about doing or creating something new until I actually see it. I have to see it to believe in order to believe that it is possible.

How often have you heard that? It's something we've all heard countless times!

But if this statement is right, how did we progress from living in caves to the world that we inhabit today? Who ever took the next step forward? Someone at every stage of civilization's advancement has had the courage to throw away that line and go do what they believed could and should be done.

Let me contrast this with a famous statement that has defined Napoleon Hill as a successful author, leader and motivator:

"Whatever the mind can conceive, and believe, it can achieve."

What Hill says allows for a space of hope or potential. If we don't think there is anything worth doing better and achieving tomorrow, why would we bother today?

We All Have Dreams

Everyone has dreams – and I'm not just talking about what happens in our heads when we close our eyes at night. But let's talk about those too. Our brain is actually so powerful that whilst we are sleeping, it can generate different images for us to select as we choose! It's sort of like a search engine?

Sometimes our brain chooses good images, and sometimes bad. Can you recall a nice dream the next day? Many of us can to the point we can even talk about it with our friends. But we don't know how to capture it and hold it firm because our brain then throws up more options for us the following day, night or week.

It's important to remember that – regardless of our current reality – we all have the ability to choose what to do with those future insights stirred up by dreams. The 'other' kind of dreaming I want to talk about is very closely related to our sleeping dreams. And again, these may be triggered by a meeting, a story, a vision, a movie, or even from something Grandma once said in passing.

In reality, we can all see it. We can mentally create a picture of a future that we like. People who succeed in achieving what they want – be that in business, family, or every day life - do so because they are able to envision their future success (as they define it) and lock onto that image consistently. This happens over and over again not only with business people but also athletes, chefs, researchers, artists, students, and performers.

The biggest problem for most of us is how to hold onto that beacon of hope, success and pleasure that we have temporarily found in our minds. A dream is a great start but a bad beginning.

From a Thought to a Clear Picture

What Napoleon Hill was trying to tell us is that if we can see the future in our own thoughts and we can somehow hold onto to it – in other words we believe it – then (and only then) can we actually achieve it.

The process has to work in reverse from normal human logic.

Roger Bannister was the first human being in the world recorded to have run a mile in less than four minutes. He achieved this amazing feat on May 6th 1954 and officially became the world's first sub four-minute miler. Within a matter of weeks more than a dozen other runners had also achieved the same milestone. But how could it be that it took mankind until the mid-20th Century to achieve this for the first time, but then immediately after, many others were capable of the same thing? Because now that it had been achieved, they no longer had to be convinced that it was even possible. They could see it! It no longer required imagination or visualisation! It wasn't a dream any more. It had become real.

The problem with dreams is that they are often short lived and life is busy. Whether you were once fighting off woolly mammoths outside your cave or now doing battle with a rival competitor in business, the distractions are constant. Today's great idea rapidly becomes tomorrow's 'lost dream'. Over time, those losses add up to a point where we even lose faith in our own ability to dream any longer or believe that anything positive can happen.

It doesn't have to be that way. If only we can empower those dreams and ideas – to give them permanent wings! Unfortunately dreams don't come with a starter kit! That's where we have to cross from inspiration to perspiration.

We have to be able to take those amazing thoughts and ideas then create them into consistent pictures in our mind, which will drive our human effort and invention forward. We need to take that bad beginning and make it an empowering beacon.

The Power of a Great Vision

So far this chapter has been a lesson in psychology and philosophy! And you thought this was a business book? It's time to change all that and put it into practical use.

We have to move from a good idea and turn it into a great Vision. We must create something that motivates and empowers us every day - a picture of the future that is so strongly compelling that we cannot resist its excitement and fulfillment.

There is a common line of thinking that is espoused by many motivational gurus that if you put a picture of what you most desire on the wall (or computer screen, in your wallet, or even the toilet door) that it will feed the dream. Can you recount a story of when you or a friend posted something up somewhere, only to find it then happened? Like a nice car or holiday destination?

Are they just good luck stories? Surely, just putting a silly picture that you cut out of a magazine or pasting an image on your PC home screen page won't achieve anything?

Well I am here to say: Yes! Be mindful what you put in front of your eyes every day. Our minds are programmed to begin to accept those images as reality! Pictures are powerful. Our minds work with them. If we give a human mind a sufficiently strong and consistent picture, it drives its owner to realise that image.

In a previous chapter, I talked about the power of sight as a strong human sense that has driven the success of social media in our world. Strong visual images are very powerful. Use them! Remember that's one of the keys to branding too?

A Business Context

I believe that a strong Vision, one that is enticing and stretches the imagination too, is one of the most powerful things that an organisation can have. If you as a company (large or small) can capture a clear picture of the future that you want to achieve, keep it front of mind and visible for all your team, then amazing things will happen. The organisation will be drawn towards that goal, objective and future image. The goal seeking nature of human beings will just make it happen.

The big question is how to do it! How do we create that picture of the future in such a manner that it can pull the organisation towards a specific desired outcome?

Let's take a quick look at what a Vision should include. Whereas a Mission talks to the Why and How of the organisation, Vision should tell us What it should look like if and when the Why and How are understood and followed. The vision is what you predict the outcome will be.

A Vision:
- Defines what that organisation will look like in the future
- Showcases what it is going to achieve
- Is a compelling beacon pulling the organisation forward into a successful future

When I talk about the future, I recommend a five to ten year time horizon. You can of course update your vision as you grow.

I have said it before and I think I need to say again as we close this chapter. "If you don't know where you are going, you will end up where you are headed!"

Without a clear plan, a view of what your life, company, or organisation will look like in the future, you have little hope of achieving something significant and valuable. And life will just take you where it flows. What if that ends up being somewhere you don't really want to be?

I believe that pictures are so much more compelling, adding significant energy and excitement, than a mere statement on a wall. I encourage you to capture your dreams, thoughts and goals onto paper in a unique way that will empower you to push on a bit harder when that next challenge arises.

Use whatever media you are comfortable with to create that picture. Having an engineering background, I used a drafting and picture package on my PC – Visio. If you are a skilled artist, draw (or have it drawn for you) a picture that shows the future desired state. Don't worry if it's clunky or poorly represented to start with. Go back and improve it later. And always use lots of colour – to make things stand out (in your mind)!

Architects are great at this – long before they break ground on any project, the entire suite of drawings includes the glossy finished look of the building being contemplated, complete with people walking about downstairs or on the boardwalk, and these are always depicted on a sunny day. People even want to buy and live in the buildings before the first brick is laid.

The additional power of a picture is that if even if you only do a half good job of deploying the Vision by getting a number of Vision pictures out into the work environment, they have a powerful subconscious effect that you can never achieve by writing a wordy statement and posting it on notice boards. Every time you walk past a picture – even your Vision – your eyes and brain take another snapshot image of it to build the mental image more strongly. As that image grows in intensity, it subtly changes our patterns of behaviour and pulls us towards that image.

Make sure you get the image right, your brain cannot distinguish between right and wrong. Only you can give it that signal! It is your choice to 'tell' your mind what to focus on. This process is so powerful that if you put in the wrong signals, pictures and images, your brain will gravitate towards them, no matter what. And then very happily achieve those wrong results!

I also believe that a Vision picture or painting offers much greater potential than just a snapshot of the future. We have used ours to show our values, our core strategies and our tactical plan. In fact it looks almost like a business process map. When you look at our Vision image, there is very little about our business that is missing from the carefully created image. It is remarkably clear about what we do and what it will look like.

The key remaining question to ask is whether you are strongly communicating your future desired state. Share your Vision in your presentations, to your team, and to your customers. That's what leadership is about – taking people towards a promising future. The more you 'teach' it, the more powerful it becomes to yourself and with the group. As leaders if, and when, we cannot clearly envision and articulate the future, then it is time to turn our role over to someone else.

References

Napoleon Hill, "Whatever the mind can conceive, and believe, it can achieve." Think and Grow Rich.

SAS PARADISE RESCUED - VISION 2020

Mission Vision Passion

paradise rescued

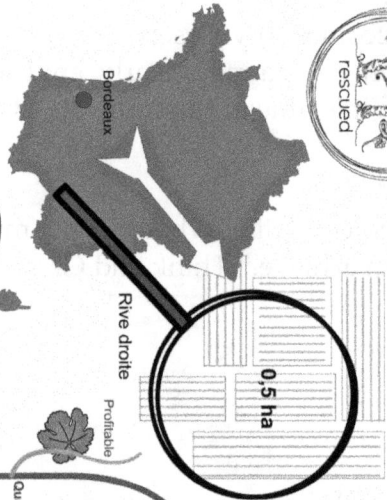

Bordeaux

Rive droite

0.5 ha

Paradise Recued Strategy

1. Good Fruit
2. Expert Vinification
3. Niche Brand Market Access

Business Structure Ownership
Key Partner Selection
Finance, Legal & IP
Strategy & Research

DMS International

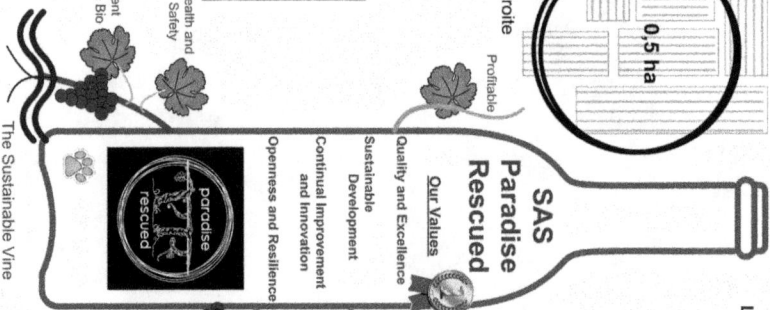

The Sustainable Vine

Environment
Elevage en Bio

Health and
Safety

Profitable

SAS Paradise Rescued

Our Values

Quality and Excellence
Sustainable Development
Continual Improvement and Innovation
Openness and Resilience

Resource efficiency

Community = people

paradise rescued

Lhoste Winery, Cardan

B1ockOne Cabernet Franc
BlockTwo MerlotCabFranc

Microcuvée

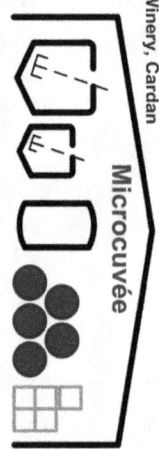

EXPORT

EXPORT

Price
Euro/b

Quality

Tactical Plan

✓ Quality not quantity
✓ Export focus
✓ Microcuvée approach
✓ Small growth steps
✓ Experiment and Learn
✓ Strong Brand Image
✓ Social Media Marketing

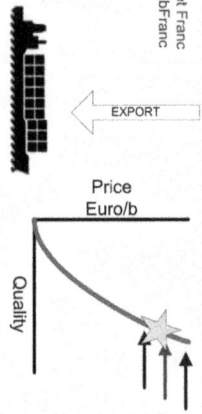

www.paradiserescued.com

#LeaderTips

1. Get a clear idea of what you want the future to look like. Convert that idea into a picture or physical image.

2. Put a copy of that image somewhere where you (and your team) can see it frequently.

3. Be bold about your future. Talk about your Vision with your team regularly. Encourage them to participate in its creation.

My three key learnings are:-

1. _____
   _____

2. _____
   _____

3. _____
   _____

My next 30 day action item is:-

_____
_____

12

Passion is the Secret Ingredient

In the previous two chapters, we looked into the why, how and what of an organisation. Hopefully by now you can see the benefit and value of having a clear Mission and Vision. Both of these are largely emotionally inspired, although the Vision part still requires some mental input to pull the dream into a coherent picture and format.

I believe that there is one other often-overlooked factor, which is critical to making an organisation succeed and excel. In the corporate world it is often categorized as one of those 'softer' touchy / feely issues that doesn't make for nice conventional, or even popular business reading. Whether we choose to talk about it or seek to ignore it, it is nevertheless a critical issue.

It is called Passion.

Passion is the magic ingredient that encompasses commitment, motivation, energy, empowerment, culture and drive. When combined with Mission and Vision, a beautiful mutually supportive trilogy comes into being.

By considering a Mission, we create the reason to develop a Vision. When we have found that Vision, we have a clear affirmation that gives us the energy and drive that reinforces to us that our Mission is achievable. That energy and drive is Passion.

With Paradise Rescued, it is often hard for our customers, fans and friends to fully appreciate and comprehend our Vision and Mission. They can sometimes see what we are achieving as projects progress and our marketing media reaches a new milestone. But one consistent theme that I hear as feedback is "Your team is so passionate". I love that.

Where does that passion come from and how do we harness it? Does it happen naturally or is it simply overdone optimism?

Let's start by approaching these questions from the other end. How many of you have recently worked with or sought service from an organisation or group of people where it all seemed 'too hard'? Where no one really looked like they wanted to be there? In fact you may have felt at the end of the meeting that you actually wished that you hadn't been there either? Was it a pleasant experience for anyone? Was any good business done or inspiring service delivered? Did anyone come away happy from the process?

If you have ever been there - as many of us have from time to time or perhaps more often - what was missing? A lack of interest, no commitment, bad attitude, low motivation, boredom, no direction? Well probably all of the above!

The Blame Game

Let's consider what many gurus and coaches may inspire us to think about... Was it really bad management or a poor workforce? It's time to grow up really and stop the blame game paradigm of 'management versus workforce'. In most cases, both are on the same payroll apparently working for the same organisation and with most likely the same organisational goals. If that organisation does not succeed, everyone loses.

Overall, it is probably fair to say that it comes down to a lack of leadership. John Maxwell says it better than most when he says "Everything rises and falls on leadership". If an organisation isn't running well, if the staff are demotivated and not delivering what the organisation wants, then it probably stems from a lack of meaning and a lack of clear direction. That's where leadership is required.

In my experience, I have rarely seen a specific group of people in a workforce who actually want to do badly and go out of their way to stuff things up. Generally, I believe most people want to achieve and perform well. So the leadership question then is about how can you get a team into an upward spiral of positive achievement as opposed to a seemingly destructive spiral downwards towards poor performance and a 'no one cares' attitude?

Human beings are goal orientated and purpose driven. This means that we love goals, we love a clear direction and we are hot-wired to achieve - providing the stimuli to encourage us, are set correctly. Humans also need a reason for doing something and we need clear direction as to what we have to achieve. It's called Mission and Vision.

The next part then is our mental connection. People can get excited about almost anything. It's incredible really. We don't actually ever play in the team, but every weekend we pay to watch our favourite sports team play and afterwards we talk about it all week long! We pay for the pleasure! We are passionate about our team. Yet when we go to work and we are paid to win, then suddenly the passion disappears? Isn't that rather hypercritical?

The key in any workplace or organisation therefore is to be able to find the passion that fires us up. And leaders very much have a role to play in achieving that aim. Leaders are responsible for setting the work environment and culture of their organisation.

I very often use the following quote by Skip Ross. It has inspired me over time and I love using it in presentations:-
"Get a clear mental picture (of what you want), fill it with emotion and drive on".

Self-Reflection Time

As leaders we are responsible for just that. We have to set the Vision and inspire our organisation to believe that it is doable, credible and worthy of everyone's effort. Whether we have ownership in the organisation or are paid to be in a leadership role, we must lead like it's our own.

If you can't find any passion and it looks like a job, guess what, you are in the wrong place. No one can give you that passion; it's not an inoculation that you get at the doctor's surgery. You have to find it from inside of you.

One of the amazing benefits of having sat down, thought, strategized and developed our Mission and Vision and then discussing it with our Paradise Rescued team plus clients on countless occasions, the passion grows with every one of these 'sharing' moments. That's both my passion and theirs.

I posed some questions for the reader to reflect on at the start of this chapter. Self reflection as a leader is an important process for which there isn't room in this book – another, maybe? I hope so! I raised the question about whether all of this leadership thinking was simply 'overdone optimism'. I need to tackle this issue with you head-on.

If you are leading an organisation or you are the business owner or maybe the founder, you must have a clear picture of what you want to achieve in the future. In this book, we have called it a Vision. If your picture of the future looks pretty similar to what is happening right now, then you need a wake up call. Or you and your organisation will be in for a fall!

If you are not optimistic about what you can achieve with and for that organisation in the future, you are not doing what a leader is there to do. Whilst change should not be made to happen just for the sake of it, believing that today's paradigm or status-quo will be good enough in the future is delusionary. A number of war-weary sometimes cynical or worn-out leaders play their 'realism' card at this stage. You have heard the phrases "Been there, tried that", "That will never work here."

I have been accused many times of being a 'Pollyanna optimist'. I have also tried being a 'miserable whatever'. Management requires risk avoidance and control. Leadership is creating the future. What we need more of is realistic thoughtful (tough-minded) optimists and a change of heart from the pessimists!

Be bold with your Mission, Vision and Passion. Seek out the pitfalls and build back-up strategies together with contingency plans. The road from now to your future Vision will not be a straight one. It requires adjustments, change and persistence. Tough-minded optimism is the way forward into uncertainty.

Being Authentic

And here is one other key point about Passion that is critical to what you achieve from your organisation. I have personally listened to dozens of managers over the years telling me, (or teams that I have led), what we should be doing or how we should be acting. And then they act in a manner, which is completely contradictory to what they request of everyone else. Fake passion is the most easily betrayed lack of integrity in the leadership book.

The so-called "Law of Attraction" can be characterised by a number of different 'one line 'statements. One of the most popular is "You get what you give". I prefer another approach, which is less often heard: "You get what you are". If you have a challenging Mission, a huge Vision but no Passion – guess what your followers will look like. I am not advocating that you go out and hire a bunch of clones – you most definitely need a range of personality types that positively compliment your own style. But Passion is not a personality type, it's an internal decision; it's an attitude driven by your belief.

If you have great passion, most likely your followers will too. And great things will happen. When you start to tap into their strengths and energy, you will find the knowledge and brilliance where yours is lacking. Even as founders, we can't be experts at everything! In the words of Simon Sinek: "The courage of leadership is giving others the chance to succeed even though you bear the responsibility for the results."

I love it when people listen to our story and say "David, you've got so much passion". That's one of the greatest compliments that I can receive. I don't have to think about it, I never have to fake it; it's 200% genuine. And it extends to those who work with us. From our international marketing partners, to our neighbours at Cardan who come and help us at harvest time, it is us, Paradise Rescued, and it's for real.

References

John C. Maxwell "Everything rises and falls on leadership" johnmaxwell.com

Skip Ross "Get a clear mental picture (of what you want), fill it with emotion and hold on" Say YES to your Potential.

#LeaderTips

1. Mission Vision Passion. Remember this tag line. Etch it into your mind. Teach it. Live it.

2. Lead like a Founder. Create a culture of passion within your business. And as the leader, be the passionate champion.

My three key learnings are:-

1. _____
   _____

2. _____
   _____

3. _____
   _____

My next 30 day action item is:-

_____
_____

13

The Prize
and the Price

It was early August 2013 and I was going through my volume of emails. As always there is a fair amount of stuff that no matter how carefully you tune your spam filter, it gets through! There in the middle of the last of so many from the previous day was one from the Stevie Awards – entitled 'Winners List Congratulations'. As I guessed it wouldn't be targeted at us, I calmly transferred it to the Delete Folder and thought no more about it.

Back at the end of the previous April, some of our Social Media Professionals network colleagues suggested that I fill out an entry form for the annual Stevie Awards, one of the world's foremost marketing, media and business awards series. I thanked them for their kindness and quickly moved on. Their requests persisted to the point where I felt compelled to at least have a look at what they were all about. I opened their web site, reviewed the list of categories and easily convinced myself that I was in the wrong place!

Their persistent requests went one step further and at the last minute I compiled the required entry form, business history plus story, paid the necessary fees and hit the Enter button. Job done – thank you.

In mid August I got another email from the Stevie Awards with a similar heading to the one I seen two weeks before. This was getting silly now and I needed to reply to someone to ensure that our name was removed from the mail list. I skimmed the email through which told me that SAS Paradise Rescued of Cardan Bordeaux France was the winner of a Silver Stevie 2013.

I now knew that this had to be a joke so I politely but firmly replied and asked them to kindly review their mail lists and ensure the letter was sent to the rightful winner.

Next day the email came back:-

"Dear Mr. Stannard,

SAS Paradise Rescued is indeed the winner of an International Silver Stevie for Best New Business 2013. Congratulations.

We hope you can join us for the Awards Gala celebration in Barcelona in early October.

Kind regards, The Stevie Awards"

I quickly went back to their web site to confirm that indeed our name was there. It was no hoax. We had won a Silver Stevie! How unreal but amazing that one of the smallest vineyards in Bordeaux, with an Australian parentage, could achieve this.

Happily the harvest time in Bordeaux that vintage coincided with the planned Awards Gala night and I was able to make the 5 hour drive down to Barcelona to collect the Silver Stevie along with David2. I also got a crash course in 30-second acceptance speech delivery as well as an unrehearsed up-close and very personal encounter with film cameras in the media zone. David2 and I returned to Cardan to proudly share the Award with our team and Monika, our US marketing partner who had flown over to meet us at the winery. Having saved a near disastrous vintage from rain, grey rot and the worst recorded Bordeaux season in 25 years, it felt like a pretty good couple of weeks at the office!

Even as I write these pages now, I still struggle to accept the reality of what had occurred. Strangely enough, I seem to be the only person who hasn't fully come to terms with the success. Why? Because it just seems like there is still so much more we can do and achieve. We are really just beginning. I can therefore confidently say that one unexpected win will not be going to my head any time soon.

But what can we share from this funny story of humble unexpected success? Always remember: success leaves clues.

Paradise Rescued has a passion and a vision of excellence in what we do. Our aim is to create a niche brand business that will organically produce top class Bordeaux red wine and export it to the world. And in so doing we will sustain the rural culture and heritage of our village in Cardan.

To achieve what we have done so far is both a reflection of the talent of our team, the skill of all those experts who have kindly shared their brilliance and our ability somehow to pull all that together into a cohesive micro business. I could not have achieved a tenth of what our team has done, often working at different ends of the project. Vine management and social media marketing are two worlds apart – trust me!

I have never been big into control – it's not me and I believe strongly in delegating responsibility and letting people take up that empowerment. Running your business requires a careful balance to ensure you have sufficient control at the right moments without having everything delegate upwards for the simplest of activities. With customers and market support in the USA and Australia, financial / legal control in two continents, marketing and media advice all over the world, a unique approach is required.

In many ways I am glad that most of my education and training was not in the wine making and marketing business. It has required me to ask and learn as well as challenge many of the current industry paradigms. I encourage you all to read, learn and listen a lot, then go and try a few things and repeat the process. Never stop learning something new about the business you are in. Doubtless it won't look the same in six months or six years. The day we stop learning, is the day that we start decaying!

Smart Business People Get Help

Small business leaders are solution focused. If you're smart, then even if you don't have all the answers you won't let yesterday's paradigms and blockers ensnare you into the same thinking. When you need help, go and find it. We have to be very careful that we don't fall into the trap of thinking that just because we are the boss and that 'we are paid the big bucks', that we actually know everything! This notion is ridiculous. If there is something that you are not good at, something which is not a core part of your company strategy or something that you need to learn but currently don't have the skills, then go get help. Now!

Here are two examples from my business. When we set out, I made some tentative efforts at label designs for our bottle, which for me conveyed what we were trying to achieve at the start. When a contact of a contact saw them, he asked a couple of questions about where we wanted to get to as a business and therefore how critical would be the role of our design and branding. He immediately sensed that we wanted to be well above average but were clueless back then about graphic design, branding and visual manifestation. As we have previously mentioned and praised, the contact we followed up with has led us to creating a unique luxury brand and registered trademark.

A second example. I have messed up more sets of accounts than I care to remember. I am neither a trained bookkeeper nor accountant, and I do not particularly like that kind of work as it gets in the way of the creative side of developing our business and customer relationship time. Of course, we have legal and financial obligations, which must be met in full as a registered business. We decided to find a great bookkeeper who takes care of everything and now leads our discussions with the accountants.

But when you do decide to seek help you need, be careful of the 'contractor' trap. In a world of ever increasing work flexibility, we are tempted to 'go hire a contractor' to fix that something that we can't or don't like to do. So what do we do? More often than not we look for the cheapest opportunity that will apparently deliver the best result at the exact second that we request it. You get what you pay for, right? If you want a partner who may go the extra mile for you and offer you that extra special difference that can take your business to the next level, then be prepared to reciprocate with your approach, timing and price. You may be surprised how much more you get back over and above what you pay for.

Ets. Leveque et Fils are our winery equipment, spares and materials supplier located in our local town of Cadillac, five kilometres from Cardan. They are a medium size successful family business and everyday their warehouse store shop is bustling with customers. We clearly are not their number 1 by total sales. However they supply us a level of service that is exceptional. When we choose to change the winery configuration mentioned in Chapter 2 above, Leveque supplied the new micro-vats. But also as part of their service, they took back the old large single vat, found a purchaser and credited the sale against our account.

Whenever I visit their premises to talk with our account manager, the owner quickly gets to hear of my arrival and personally passes by – with a coffee at the ready – to welcome me. When I have a key winery or vineyard issue to resolve, guess where I go and who will get that business? Are they the cheapest? I doubt it but we get more value in return that we can ever imagine or the balance sheet can show.

Managers and Leaders

Play to your strengths. In a management world we are taught to work on improving our weaknesses. And in our annual appraisals, we spend more time reviewing the things we aren't so good at. However, in a leadership world, we learn to focus on extending our strengths, secure in the knowledge that as we grow them, those areas where we don't excel will grow too. A rising tide lifts all ships.

Which brings me to a final subject; please understand the differences between leadership and management. The two words frequently become interchanged as if they mean the same thing. They don't. Sure, they get blurred from time to time and even overlap. But as principles they start from a long way apart!

Management is focused on control, efficiency, things and risk aversion. Leadership is about, direction, communication, people and effectiveness.

Yes, there is a huge amount I could write on both topics - this sounds like the basis for that next book? Both have their place and both are necessary in running a big or small business. Both can sometimes be delegated but neither can be abdicated. There is always 'fun stuff to do' in running a business but sometimes there are tough matters to be taken care of too.

#LeaderTips

1. Don't do it all yourself. No one can be good at everything. Get good help and reward / recognise it well (money not always required!)

2. But most of all, believe in yourself. Share your Mission, Vision and Passion with your team. And then believe in them too!

3. Don't let the success go to your head. Keep learning. Share what you learn. Enjoy the journey.

My three key learnings are:-

1. _____
   _____

2. _____
   _____

3. _____
   _____

My next 30 day action item is:-

_____
_____

14

Il n'est pas
le Terroir!

Il n'est pas le Terroir? No. Definitely not! It's not about the Terroir. It's not about the Dirt.

In From Cabbage Patch to Cabernet Franc, we explored the seemingly French and unique concept of Terroir. And the mysterious almost secret notion that: unless you have a block of vines in exactly the right location, with the perfect combination of:-

- soil / geology,
- topography / geography and
- microclimate,

you will be unable to create a wine of any significant quality.

In classical French wine mythology, one is given to believe at a large number of Château cellar doors in Bordeaux that the Terroir dictates everything. If one wine is better than a neighbour's, it is supposedly because of some magical intangible quality related to those three factors of that vineyard that make up the classical definition of Terroir.

On one side of the road, the owner can label his wine as a Grand Cru Classé of St. Emilion and just ten metres away, it would be a humble Appellation Bordeaux. How come a simple hop, skip and a jump over a seemingly invisible geological barrier and short distance created such a huge difference? And then why on a steep rocky vineyard above the River Rhône, better suited to mountain goats than vignerons, are some of the world's great Syrah wines created?

I am personally fascinated by both the assumed concept of the word Terroir and, consequently, therefore the potential reality of its assertion. In other words: is it true? Whilst we never set out to make mediocre wine, the journey towards excellence has taught our team and myself a good many excellent lessons. And on that journey of discovery, not only have we come closer to unraveling the secrets hidden in Terroir but we have learnt some more about the make-up of human beings too!

To me as an engineer, none of this Terroir 'stuff' made any logical sense. It just did not fundamentally add up. It has taken me a few years, much discussion (argument?), observation and our own personal experience to unravel the mystery.

We have discovered that Terroir includes another factor. It has a fourth dimension. A human dimension. Both technically and psychologically. However it is an important one to learn; not only as wine makers but also as business owners, leaders and human beings in general. Hence I have saved the human dimension until last in this book as a way of summarising this book but also opening the door for a potential next book at some time in the future.

Please don't misunderstand me! St. Emilion is (and I hope always will be) a beautiful, brilliant, amazing Bordeaux wine centre and a beacon of viticultural excellence. Some of the world's greatest Merlot and Merlot / Cabernet Franc blends are produced there. It has a good number of advantages from the perspective of growing amazing grapes and producing outstanding wines. As a location, it pretty well ticks all traditional boxes for a good Terroir that we highlighted at the start of this chapter. But at the same time, even in the middle of that beautiful metropolis, there are a number of cabbage patches making some very average wines.

This shouldn't be a great shock to anyone, even if it isn't a fact that top wine growers from that area will ever talk about with you! So although there is a roughly equal advantage for anyone growing grapes in that area, some people/organisations manage to excel at their craft a lot better than others and have done so over a long period of time. Hence the traditional values of Terroir do not completely explain nor automatically guarantee viticultural success and a Grand Cru Classé appellation on their labels.

The difference therefore is not whether one vineyard is capable of making better wine than its neighbour as the overall differences in soil, geography and microclimate are small. It becomes a question of whether one Château is using both the human skills necessary to make great wine AND combining it with the belief and vision that it can achieve that goal. Finally of course, a plan and activity level is needed to actually make it all happen.

Humans beings are human beings. Some have found the secrets to making and sustaining that mini-region in a successful manner whereas others have been unwilling (or unable) to use the natural advantage given to them. But the difference - as so often – comes down to people. Their skills, knowledge, vision, passion and expertise. Time and time again, we hear the success stories of teams, organisations and companies that value and benefit from what their people have been able to achieve when they are motivated, passionate and well led. Over and over we learn about groups of people who have learned to overcome significant disadvantage and succeed compared to others who seemingly started from a more beneficial position.

This tells us a lot about the psychology of people and the apparent problem of blindly believing in "traditional Terroir". Maybe all people operate their lives under a Terroir system, where we create our own boundaries and beliefs about what we are capable of, what we can achieve and (worse still) what we can't realise?

Was Paradise Rescued just lucky? I don't believe so. If I look at our Hourcat Sud vineyard block, the home of BlockOne Cabernet Franc (our original cabbage patch) and do a very honest self assessment of it's apparent traditional Terroir qualities, it really does not tick many of the boxes! The geology and soil is passable but most certainly not in the same league the same as St. Emilion! The topography aspect of the vineyard block slopes towards the west and the vines run east west; ideally one would prefer a south facing slope and north south row orientation. The climate in Bordeaux is generally positive but our microclimate is humid with intermittent winter surface water logging. It also has challenging, ever changing drainage conditions. Hardly ideal growing conditions?

Whilst as it's owner, I didn't initially have sufficient self-belief that the Hourcat Sud block could produce a superior varietal Cabernet Franc, our team simply ignored those beliefs. They acted as if it were a piece of St. Emilion. They applied their viticultural science and knowledge in the same way as if it were a vineyard located around that magical town. The results speak (taste?) for themselves. Although they have been applying and consolidating their skills for only a relatively short time, their achievements are most definitely not just 'beginners luck'!

They have clearly proven that there is a fourth dimension to Terroir – the human dimension.

And they are using it to the maximum of their ability. They are continuing to nurture that experience and feed it back into the vineyard, the soil and our wine. I very much doubt – even in a good vintage – that our team has peaked yet. There is more to come as they build knowledge, improve their techniques further and consolidate the full dimension of our Terroir.

This speaks volumes about life and human beings in general. It gives us all a great lesson to appreciate. It tells us so much about human potential. Our vineyard operations team never saw the location of our vineyard as a disadvantage – they simply set out to maximise its potential by passionately applying the best technology and knowledge possible for that piece of vineyard. They effectively set out to emulate what the region of St. Emilion does. In fact, they transported St. Emilion to Cardan. The so-called 'facts' didn't matter. They had a real Mission, a clear Vision and they added the Passion with the capacity of their human brilliance.

And they are succeeding – big time!

As their leader, I can now appreciate the lesson they taught me.

As people, we too often limit our potential by our thinking and the self-imposed limitations that we place on ourselves – or the limitations that we allow other people to place on us. We all create our own glass ceilings that we apply to our lives. We all "find comfort" in having a natural boundary beyond which it appears too unsafe to go and will require huge personal development to overcome.

Henry Ford's famous quote immediately comes into mind: "Whether you think you can, or you think you can't - you're absolutely right." I fully agree.

As business leaders, it is no different. When we believe that we can create something of value, we find new ways and solutions. People who change the paradigms of the world do so because they have a clear picture of what they want to achieve, why they are doing it and how they propose to go about it.

As soon as we lose that Mission, Vision or Passion, we are lost!

As we gain more wisdom and lose more youth, that ability and determination to extend ourselves beyond our (self imposed) boundaries diminishes. The natural longer term outcome of the full process is individual sadness and a resignation to simply endure the rest of what life has to throw at us.

Returning full circle to the vineyards, there is another similar business (but human) lesson. A piece of vineyard, a good business, great technology or superb assets may be your birthright but it is not a right of passage to continued success. Success comes through people and the ongoing accumulation of knowledge and its application.

If you want a vineyard or business to continue to sustain its position of excellence or to grow to a new previously unattainable level of performance, you need to make investments. A good part of that investment is in human capital; its development and the capture of greater knowledge.

Our Hourcat Sud Cabernet Franc vineyard is a good metaphor. We are starting to reveal its potential as a viticultural asset. To achieve what we have done so far and go to the next levels that we have planned, has required and will continue to require, a good level of ongoing human skill, passion and knowledge.

If you want great results, you need to find and / or develop great people.

Many of the world's most successful and enduring companies are medium sized private niche operations. At first glance, this might make no sense but as you reflect back on the contents of this book, you can see that many of the human factors we have discussed are consistently present and being pursued day in, day out. We talk about "People making the Difference". But do we really believe it and make it happen?

Life, business or a vineyard; it's not about what you are given. It's what you do with that opportunity that counts.

That's why I close with these words: "Il n'est pas le Terroir". It's not about the dirt !

#LeaderTips – the Time for Action

In Chapter 1, we introduced our '#LeaderTips' section comprising a few helpful hints at the end of each chapter. Almost everything that we have used or developed in Paradise Rescued didn't start with us! I am not telling you to go out and misuse a patent or deliberately steal someone else's technology. I am encouraging you however to keep your eyes / ears open and when you see things that work successfully in another business to then apply those same or similar ideas in your own business. This is very much part of what I call innovation. The internet is wide open with ideas from all over the world. Take a couple and integrate them into improving your business.

And finally, above all else THINK! Think about what you are doing. Apply yourself mentally to your business. Don't just work IN it, work ON it! Today's model of success is unlikely to be the one that succeeds tomorrow. Adapt, innovate and plan the future that you want. Think about why, how and what you want to achieve. And commit 200% to it.

Mission Vision Passion! Grow yourself – Grow your Business. You can do it!

References

Henry Ford "whether you think you can or whether you think you can't – you're absolutely right."

#LeaderTips

1. Check your boundaries. Revisit your Vision. What self-imposed limitations do you have? Are they real? Challenge every one of them? Set a plan to expand your personal and / or business horizons. Get rid of your glass ceilings. #YouCanDoIt

2. Most likely your business or organisation has some key people, skills, knowledge and expertise. Develop a plan to grow, nurture and maintain that base of human learning. Continue to build your company through your people. Take them on the growth journey with you.

My three key learnings are:-

1. _____
   _____

2. _____
   _____

3. _____
   _____

My next 30 day action item is:-

_____
_____

Acknowledgements

Team Paradise Rescued. We learnt this all together. In the vineyard, winery, on social media, the presentations and one on one time together.

Tricia Wiles, Sweet Graphic Design for the brilliance that is the Paradise Rescued logo and amazing international micro brand. And of course for the book design!

Dixie Maria Carlton for courageously taking on the goal of producing this second book for me and fine tuning our marketing and branding. As well as changing my thinking and self belief.

Zoe Wyatt, Social Media ShortCut for the social media marketing wisdom and help from around the world. What I wrote in here started from you.

Victor Caune and Dick Sandner for your pre-reviews of the book, feedback and opening the doors to the business market potential of our book contents.

Cheers Monika Elling, Foundations Marketing Group, New York for hanging in there with us throughout and holding open that window to bring our product to market in the USA. We are getting there.

Our business partners and loyal advisors of Paradise for your belief, encouragement and patience.

Similarly to my friends and family for the non-stop support.

To our Club Paradise Rescued customers, fans and brand ambassadors. Many of the great ideas and thoughts in this book come from your feedback, emails and tweets. Please keep those thoughts coming; they are the engine of our development.

My family who stand by no matter how long the days and challenging the situations.

And of course I will never forget why we do what we do – our Mission: to sustain the rural heritage of our beautiful village and commune of Cardan, Bordeaux. To our village, neighbours and community – thank you.

About David Stannard

David Stannard is the Founder and Owner Director of Paradise Rescued.

David was born and educated in England, qualifying as a Chemical Engineer from Birmingham University in 1980. He has had a successful leadership career in the petrochemical industry in the UK, Netherlands and Australia covering more than 30 years.

In 2010, as a tsunami of new housing threatened to wipe away the rural heritage of the village of Cardan in the vineyards of Bordeaux France, Paradise Rescued was founded. Working closely with the community, David successfully brought together a dedicated and passionate team to manage the vineyard, winery, export marketing and business brand development.

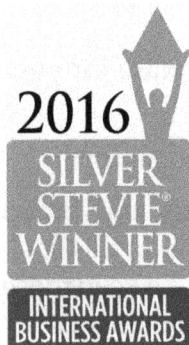

Today Paradise Rescued produces and exports organically produced red wines to Australia and the USA and is internationally recognised as a niche ultra premium micro wine brand. Paradise Rescued was the Silver Stevie International Business Award winner 2013 for Best New Company and the Silver Stevie International Business Award Winner2016 for Small Budget Marketing Campaign (<$5M). The Paradise Rescued Cabernet Franc 2010 vintage was awarded a Bronze medal at the 2016 Melbourne International Wine Show.

David and Team Paradise Rescued are highly sought after as advisers and leadership partners for other international Bordeaux wine business projects.

www.paradiserescued.com
https://www.linkedin.com/company/paradise-rescued
Twitter – ParadiseRescued
Facebook – Paradise Rescued
Instagram – ParadiseRescued
Pinterest – Paradise Rescued

www.ingramcontent.com/pod-product-compliance
Lightning Source LLC
Chambersburg PA
CBHW071704210326
41597CB00017B/2318